Intelligent Network Management and Control

SCIENCES

Networks and Communications, Field Director – Guy Pujolle

Network Management and Control, Subject Head – Francine Krief

Intelligent Network Management and Control

Intelligent Security, Multi-criteria Optimization, Cloud Computing, Internet of Vehicles, Intelligent Radio

Coordinated by
Badr Benmammar

iSTE

WILEY

First published 2020 in Great Britain and the United States by ISTE Ltd and John Wiley & Sons, Inc.

ISTE Ltd
27-37 St George's Road
London SW19 4EU
UK

www.iste.co.uk

John Wiley & Sons, Inc.
111 River Street
Hoboken, NJ 07030
USA

www.wiley.com

Library of Congress Control Number: 2020937506

British Library Cataloguing-in-Publication Data
A CIP record for this book is available from the British Library
ISBN 978-1-78945-008-8

ERC code:
PE7 Systems and Communication Engineering
PE7_1 Control engineering
PE7_8 Networks (communication networks, sensor networks, networks of robots, etc.)

MIX
Paper from responsible sources
FSC® C013604

Contents

Introduction

Badr Benmammar
Abou Bekr Belkaid University, Tlemcen, Algeria

Computer network management and control previously involved mainly purely technical tasks of network equipment maintenance, with the goal of ensuring their proper operation and further development.

Due to the current emergence of computer networks and the development of a steadily growing number of applications able to operate on a network (and more generally on the Internet), the management and control of a computer network can no longer do without artificial intelligence throughout its stages.

This is what our introductory book to intelligent management and control of computer networks endeavors to prove. Our aim is to present the use of artificial intelligence in networks through their intelligent control.

The main objective of artificial intelligence is to design systems that are able to replicate human behavior in its reasoning activities. Defining artificial intelligence is nevertheless not a simple task. The extent of the field is such that it is impossible to narrow it down to a specific field of research.

According to the definition given by one of its creators, Marvin Lee Minsky, artificial intelligence is "the construction of computer programs that engage in tasks that are currently more satisfactorily performed by human beings because they require high-level mental processes such as: perceptual learning, memory organization and critical reasoning".

Intelligent Network Management and Control,

In computer networks, artificial intelligence applications relate to several domains, such as intelligent radio communications, new communication architectures, cloud computing, network optimization and security.

This book addresses topical issues that are mainly related to intelligent security for computer networks, deployment of security services in software-defined networking (SDN), optimization of networks by means of artificial intelligence techniques and multiple criteria optimization methods for the selection of networks in a heterogeneous environment. The book deals also with the selection of cloud computing services, intelligent computation offloading in the context of mobile cloud computing, smart management of resources in a smart grid-cloud system for better energy efficiency, Internet of vehicles (IoV), relying on its new architectures, artificial intelligence application in cognitive radio networks and finally the contribution of intelligent radio to addressing the road communication needs of autonomous vehicles.

The various subjects dealt with in this book are organized in parts, each of which contains two chapters. This structure is intended to make it easier for the reader to comprehend the contribution of artificial intelligence to each specific field.

PART 1

AI and Network Security

1

Intelligent Security of Computer Networks

Abderrazaq SEMMOUD and Badr BENMAMMAR
Abou Bekr Belkaid University, Tlemcen, Algeria

1.1. Introduction

Artificial intelligence (AI) and machine learning have rapidly progressed in recent years, facilitating the development of a broad range of applications. For example, AI is an essential component of widely used technologies such as automatic speech recognition, machine translation, spam filters and facial recognition. Promising technologies are currently the object of research or small-scale pilot projects, among which it is worth mentioning self-driving cars, digital assistants and drones activated by AI. Looking further into the future, advanced AI may reduce the need for human labor and improve governance quality.

A wide variety of tasks are automated using AI. Games, car driving and image classification are some of the tasks commonly studied by AI researchers. A broad set of tasks can be transformed by AI. At the very least, every task requiring human intelligence is a potential target for AI innovation. While the field of AI dates back to 1950, several years of rapid progress and growth have recently led to higher reliability. Sudden performance gains have been accomplished by researchers in a number of fields. Figure 1.1 illustrates this trend in the case of image recognition, where over the past few years AI systems have increased their performance in terms of classification accuracy from about 70% to nearly perfect classification accuracy (98%), which surpasses the human reference (95%) (Brundage *et al.* 2018).

Intelligent Network Management and Control,
coordinated by Badr BENMAMMAR.

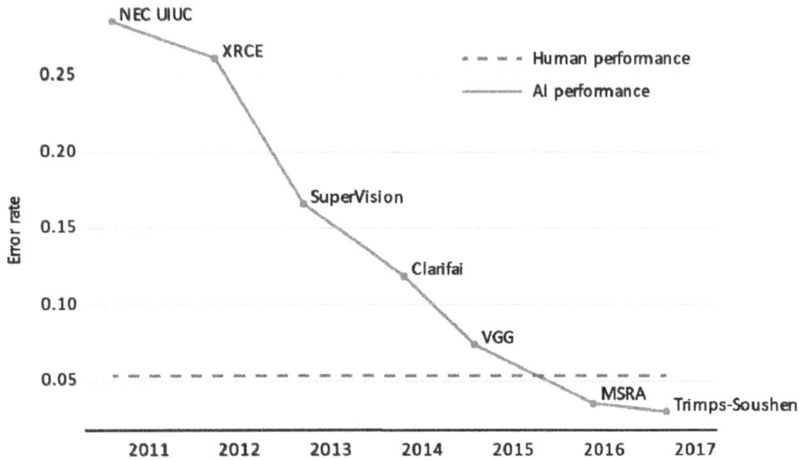

Figure 1.1. *Progress in image recognition (benchmark ImageNet), "Electronic Frontier Foundation's AI Progress Measurement" (August 2017)*

From a security perspective, a number of AI developments are worth mentioning. For example, target-face recognition and space navigation capacities are applicable to autonomous weapons systems. Similarly, image, text and voice generation possibilities could be used online to imitate other persons or influence public opinion by disseminating AI-generated content via social networks. These technical developments can also be considered early indicators of the potential of AI. Unsurprisingly, AI systems may soon qualify for an even wider range of security related tasks.

Information security is defined as the protection of computer systems against any unauthorized access, use, disruption, modification or destruction in order to provide confidentiality, integrity and availability (Peltier 2010). Information security does not refer to any particular security technology, but rather to a strategy involving persons, processes, rules and tools required in order to detect, prevent, document and mitigate current threats. With increasingly interconnected networks, security services are becoming ever more important. Connectivity is no longer an option in the commercial world, and its potential risks do not outweigh its advantages. Consequently, cybersecurity services should offer adequate protection to companies operating in a relatively open environment. Compared to classical approaches to computer security, several new hypotheses related to current computer networks should be formulated:

– modern networks are very large and further interconnected, and they are more accessible; consequently, potential attackers can easily connect and access these networks remotely;

– network interconnection increases the probability of attacks directed at large size networks such as the Internet by means of a set of widely known and open protocols.

The complexity of computer systems and applications is steadily growing. Consequently, it has become increasingly difficult to correctly analyze, secure and test computer system security. When these systems and their applications are connected to large networks, the risk of threats significantly increases. In view of adequate protection of computer networks, the deployed procedures and technologies must ensure (Khidzir *et al.* 2018):

– *confidentiality*: due to data confidentiality, only authorized users have access to sensitive information;

– *integrity*: due to data integrity, only authorized users can modify sensitive information; integrity could also ensure data authenticity;

– *availability*: due to system and data availability, authorized users have uninterrupted access to resources and important data.

The confidentiality, integrity and availability triad is a fundamental concept of information security. Each organization strives to ensure these three elements of the information system. Confidentiality prevents unauthorized disclosure of sensitive information (Kumar *et al.* 2018). Integrity prevents any unauthorized modification of information, thus ensuring information accuracy. Cryptographic hashing functions (such as SHA-1 or SHA-2) can be used to ensure data integrity. Availability prevents loss of access to resources and information (Kumar *et al.* 2018).

1.2. AI in the service of cybersecurity

AI systems are generally efficient, being less time and money-consuming than a human being when fulfilling a given task. AI systems are also evolutionary, as their computation power enables the completion of far more tasks in the same amount of time. For example, a typical facial recognition system is both efficient and evolutionary; once developed, it can be applied to numerous camera flows with a significantly lower cost than that of human analysts employed to perform a similar job. This explains why cybersecurity experts are seriously looking into AI and its potential contribution to mitigating certain problems. As an example, machine learning used by many AI algorithms can help detect malware, which are increasingly difficult to identify and isolate due to their growing capacity to adapt to traditional security solutions (Veiga 2018).

Capgemini Research Institute has conducted a survey of 850 managers of seven large industrial companies: among the top management members included in this survey, 20% are information systems managers and 10% are responsible for information systems security. Companies headquartered in France, Germany, United Kingdom, the United States, Australia, India and Italy are mentioned in the report (Capgemini Research Institute 2019). Capgemini noted that, as digital companies develop, their cyberattack risk increases exponentially. It has been noted that 21% of companies declared one cybersecurity breach experience leading to unauthorized access in 2018. The price paid by companies for cybersecurity breaches is heavy (20% declared losses of over 50 million dollars). According to this survey, 69% of the companies estimate a need for AI to counteract cyberattacks. The majority of telecommunications companies (80%) declared that they relied on AI to identify the threats and counteract the attacks. According to the Capgemini report, the telecommunications sector declared the highest losses of over 50 million dollars, which led to AI being considered a priority in counteracting the costly breaches in this sector. Understandably, consumer goods sellers (78%) and banks (75%) came second and third, respectively, in this ranking, as these sectors increasingly rely on digital models. Companies based in the United States have as their top priority AI-based cybersecurity applications and platforms.

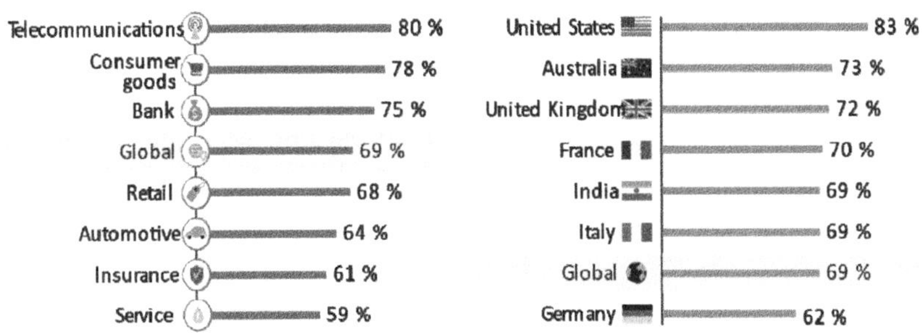

Figure 1.2. *Organizations and countries relying on artificial intelligence to identify threats and counteract attacks*

New vulnerabilities are discovered every day in the current programs, and these may infect and take control of a company's entire network. In contrast to traditional software vulnerabilities (for example, buffer memory overflow), the current intelligent systems have a certain number of vulnerabilities. This involves in particular data input causing errors in learning systems (Biggio *et al.* 2012), taking advantage of the flaws in the design of autonomous systems' objectives (Amodei *et al.* 2016) and the use of inputs designed to falsify the classification of machine learning systems (Szegedy *et al.* 2013). As these vulnerabilities show, intelligent systems may outperform humans, but their potential failures are also unrivaled.

An ideal cyberdefense would offer full protection to users, while preserving system performances. Although this ideal cyberdefense may currently seem very distant, steps could be taken toward it by rendering cyberdefense more intelligent. The idea of using AI techniques in cybersecurity is not new. Landwehr (2008) states that, at their start, computer security and AI did not seem to have much in common. Researchers in the field of AI wanted computers to do by themselves what humans were able to do, whereas the researchers in the security field tried to solve the leakages in the computer systems, which they considered vulnerable. According to Schneier (2008), "The Internet is the most complex machine ever built. We barely understand how it works, not to mention how to secure it". Given the rapid multiplication of new web applications and the increasing use of wireless networks (Barth and Mitchell 2008) and the Internet of Things, cybersecurity has become the most complex threat to society.

The need for securing web applications against attacks (such as Cross Site Scripting [XSS], Cross Site Request Forgery [CSRF] and code injection) is increasingly obvious and pressing. Over time, XSS and CSRF scripts have been used to conduct various attacks. Some of them can be interpreted as direct bypasses of the original security policy. The same security policy was similar to a simple and efficient protection, but it turned out it could be easily bypassed and certain functionalities of modern websites could be blocked. According to Crockford (2015), the security policies adopted by most browsers "block useful contents and authorize dangerous contents". These policies are currently being reviewed. However, the detection of attacks such as XSS, CSRF or code injection requires more than a simple rule, namely a context-dependent reasoning capacity.

The use of AI in cybersecurity generally involves certain smart tools and their application to intrusion detection (Ahmad *et al.* 2016; Kalaivani *et al.* 2019) or other aspects of cybersecurity (Ahlan *et al.* 2015). This approach involves the use of other AI techniques developed for problems that are entirely different from cybersecurity; this may work in certain cases, but it has inherent and strict limitations. Cybersecurity has specific needs, and meeting them requires new specifically developed AI techniques. Obviously, AI has substantially evolved in certain fields, but there is still a need for learning and developing new intelligent techniques adapted to cybersecurity. In this context, according to Landwehr (2008) one "AI branch related to computer security from its earliest age is automated reasoning, particularly when applied to programs and systems. Though the SATAN program of Dan Farmer and Wietse Venema, launched in 1995, has not yet been identified as AI, it has automated a process searching for vulnerabilities in system configurations that would require much more human efforts". Ingham *et al.* (2007) have proposed an inductive reasoning system for the protection of web applications. The works of Vigna and co-workers (Mutz *et al.* 2007; Cova *et al.* 2007, 2010; Kirdaa *et al.* 2009; Robertson *et al.* 2010) have also dealt with the protection of web applications against cyberattacks. Firewalls using deep packet inspections can be considered a sort of AI instantiation in cybersecurity. Firewalls have been part of the cyberdefense arsenal for many years. Although in most cases more sophisticated techniques (Mishra *et al.* 2011; Valentín and Malý 2014; Tekerek and Bay 2019) are also used, filtering relies on the port number. Firewalls cannot rely on the port number, as most web applications use the same port as the rest of the web traffic. Deep packet inspection is the only option enabling the identification of malware code in a legitimate application. The idea of application layer filtering of the Transmission Control Protocol/Internet Protocol (TCP/IP) model was introduced in the third generation of firewall in the 1990s. The modest success of these technologies is an indication that much more is still to be done in AI, so that it can make a significant difference in terms of cybersecurity. Nevertheless, it is worth noting that using AI in cybersecurity is not necessarily a miracle solution. For example, attacks without malware, which require no software download and dissimulate malware activities inside legitimate cloud computing services, are on the increase, and AI is not yet able to counteract these types of network breach.

1.3. AI applied to intrusion detection

Intrusion detection is defined as the process of intelligent monitoring of events occurring in a computer system or network and their analysis in search for signs of security policy breach (Bace 2000). The main objective of intrusion detection systems is to protect network availability, confidentiality and integrity. Intrusion detection systems are defined both by the method used to detect the attacks and by their location in the network. The intrusion detection system can be deployed as a network- or host-based system in order to detect the anomalies. Abusive use is detected based on the correspondence between known models of hostile activities and the database of previous attacks. These models are very effective for identifying known attacks and vulnerabilities, but less relevant in identifying new security threats. Anomaly detection looks for something rare or uncommon, applying statistical or intelligent measurements to compare the current activity to previous knowledge. Intrusion detection systems rely on the fact that they often need many data for the artificial learning algorithms. They generally require more computer resources, as several metrics are often preserved and must be updated for each system activity (Ahmad *et al.* 2016). The intrusion detection expert system (IDES) (Lunt 1993) developed by Stanford Research Institute (SRI) formulates expert knowledge on the known models of attack and vulnerabilities of the system in the form of if–then rules. The time-based inductive machine (Teng and Chen 1990) learns several sequential models to ensure the detection of anomalies in a network. Several approaches using the artificial neural networks for intrusion detection systems have been proposed (Kang and Kang 2016; Kim *et al.* 2016; Vinayakumar *et al.* 2017; Hajimirzaei and Navimipour 2019). AI-based techniques are categorized in various classes (Mukkamala and Sung 2003a; Novikov *et al.* 2006).

1.3.1. *Techniques based on decision trees*

Decision trees are powerful and widespread nonparametric learning tools used for classification and prediction problems. Their purpose is to create a model that predicts the values of the target variable, relying on a set of sequences of decision rules deduced from learning data. Rai *et al.* (2016) have developed an algorithm based on the C4.5 decision tree approach. The most relevant characteristics are selected by means of information gain and the fractional value is selected so that it renders the classifier unbiased with respect to the most frequent values. In the work of Sahu and Babu (2015), a database referred to as "Kyoto 2006+" is used for the experiments. In Kyoto 2006+, each instance is labeled as "normal" (no attack), "attack" (known attack) and "unknown attack". The Decision Tree algorithm (J48) is used to classify the packets. Experiments confirm that the generated rules operate with 97.2% accuracy. Moon *et al.* (2017) proposed an intrusion detection system based on decision trees using packet behavior analysis to detect the attacks. Peng *et al.* (2018) proposed a technique that involves a preprocessing for data digitization, followed by their normalization, in order to improve detection efficiency. Then a method based on decision trees is used.

1.3.2. *Techniques based on data exploration*

Data exploration aims to eliminate the manual elements used for the design of intrusion detection systems. Various data exploration techniques have been developed and widely used. The main data exploration techniques are presented in the following sections.

1.3.2.1. *Fuzzy logic*

Fuzzy logic has been used in the field of computer networks security, particularly for intrusion detection (Idris and Shanmugam 2005; Shanmugavadivu and Nagarajan 2011; Balan *et al.* 2015; Kudłacik *et al.* 2016; Sai Satyanarayana Reddy *et al.* 2019), for two main reasons. First, several quantitative parameters used in the context of intrusion detection, for example processor use time and connection interval, can be potentially considered as fuzzy variables. Second, the security concept is itself fuzzy. To put it differently, the fuzzy concept helps in preventing a sharp distinction between normal and abnormal behaviors. Kudłacik *et al.* (2016) have applied fuzzy logic for intrusion detection. The proposed solution analyzes the user activity over a relatively short period of time, creating a local user profile. A more in-depth analysis involves the creation of a more general structure based on a defined number of local user profiles, known as a "fuzzy profile". The fuzzy profile represents the behavior of the computer system user. Fuzzy profiles are directly used in order to detect user behavior anomalies, and therefore potential intrusions. Idris and Shanmugam (2005) proposed a modified FIRE system. It is a mechanism for the automation of the fuzzy rule generation process and the reduction of human intervention making use of AI techniques.

1.3.2.2. *Genetic algorithms*

Genetic algorithms are techniques derived from genetics and natural evolution, which have been used to find approximate solutions to optimization and search problems. The main advantages of genetic algorithms are their flexibility and robustness as global search method. As for drawbacks, they are computationally time-consuming, as they handle several solutions simultaneously. Genetic algorithms have been used in various manners in the field of intrusion detection (Hoque *et al.* 2012; Aslahi-Shahri *et al.* 2016; Hamamoto *et al.* 2018). Hoque *et al.* (2012) presented an intrusion detection system using a genetic algorithm to effectively detect anomalies in the network. Aslahi-Shahri *et al.* (2016) proposed a hybrid method that uses support vector machines and genetic algorithms for intrusion detection. The results indicate that this algorithm can reach a 97.3% true positive rate and a 1.7% false positive rate.

1.3.3. ***Rule-based techniques***

Rule-based techniques (Li *et al.* 2010; Yang *et al.* 2013) generally involve the application of a set of association rules for data classification. In this context, if a rule stipulates that *if event X occurs, then event Y is likely to occur*, events X and Y can be described as sets of pairs (*variable, value*). The advantage of using rules is that they tend to be simple and intuitive, unstructured and less rigid. Nevertheless, a drawback is that rules are difficult to preserve and, in certain cases, inadequate for the representation of various types of information.

Turner *et al.* (2016) developed an algorithm for monitoring the enabled/disabled state of the rules of an intrusion detection system based on signatures. The algorithm is implemented in Python and runs on Snort (Roesch 1999). Agarwal and Joshi (2000) proposed a general framework in two stages for learning a rule-based model (PNrule) in order to learn classifier models on a set of data. They extensively used various distributions of classes in the learning data. The KDD Cups database was used for learning and testing their system.

1.3.4. *Machine learning-based techniques*

Machine learning can be defined as the capacity of a program to learn and improve the performances of a series of tasks in time. Machine learning techniques focus on the creation of a system model that improves its performances relying on the previous results. Furthermore, it can be said that machine learning–based systems have the capacity to handle the execution strategy depending on the new inputs. The main machine learning techniques are presented in the following sections.

1.3.4.1. *Artificial neural networks*

Artificial neural networks learn to predict the behavior of various system users. If correctly designed and implemented, neural networks can potentially solve several problems encountered by rule-based approaches. The main advantage of neural networks is their tolerance to inaccurate data and uncertain information and their capacity to deduce solutions without previous knowledge on data regularities. Cunningham and Lippmann (2000) of MIT Lincoln Laboratory conducted a number of tests using neural networks. The system searched for attack-specific key words specific in the network traffic. In Ponkarthika and Saraswathy (2018), a model of intrusion detection system is explored as a function of deep learning. Long–short term memory (LSTM) architecture was applied to a recurrent neural network for the learning of an intrusion detection system using the KDD Cup 1999 dataset.

1.3.4.2. *Bayesian networks*

A Bayesian network is a probabilistic graphical model representing a set of random variables in the form of an acyclic oriented graph. This technique is generally used for intrusion detection in combination with statistical diagrams. It has several advantages, notably the capacity to code the interdependences between variables and to predict events, as well as the possibility of integrating both previous knowledge and previous data (Heckerman 2008). Its major drawback is that results are comparable to statistical techniques, but this requires additional computation efforts. Kruegel *et al.* (2003) proposed a multisensor fusion approach using a Bayesian network–based classifier for the classification and cancellation of false alarms, according to which the outputs of various sensors of the intrusion detection system are aggregated to generate a single alarm. Han *et al.* (2015) proposed an intrusion detection algorithm based on Bayesian networks relying on the analysis into main components. The authors calculate the characteristic data value of the attack on the original network, and then extract the main properties by analysis into main components.

1.3.4.3. *Markov chains*

A Markov chain is a random process related to a finite number of states, with memoryless transition probabilities. During the learning phase, probabilities associated with transitions are estimated from the normal behavior of the target system. Detection of anomalies is then achieved by comparing the anomaly score obtained for the sequences observed at a fixed threshold. In the case of a hidden Markov model (Hu *et al.* 2009; Zegeye *et al.* 2018; Liang *et al.* 2019), the system we are interested in is assumed to be a Markov process in which states and transitions are masked. In the literature, several methods have been presented for solving the intrusion detection problem by inspecting the packet headers. Mahoney and Chan (2001) experimented with anomaly detection on DARPA network data by comparing the header fields of the network packet. Several systems use the Markov model for intrusion detection: PHAD (Packet Header Anomaly Detector) (Mahoney and Chan 2001), LERAD (Learning Rules for Anomaly Detection) (Mahoney and Chan 2002a) and ALAD (Application Layer Anomaly Detector) (Mahoney and Chan 2002b). In the book of Zegeye *et al.* (2018), an intrusion detection system using the hidden Markov model is proposed. The phase of network traffic analysis involves characteristic extraction techniques, reduction of dimensions and vector quantization, which plays an important role in large sets of data, as the amount of data transmitted increases every day. Model performances with respect to the KDD 99 dataset indicate an accuracy above 99%.

1.3.4.4. *Support-vector machines*

The support-vector machine is a technique used for solving various learning, classification and prediction problems. The support-vector machine was employed in an implementation of the structural risk minimization (SRM) principle of Vapnik (1998), which minimizes the generalization error, in the sense of true error on unseen examples. The basic support-vector machine addresses problems with two classes, in which data are separated by a hyperplane defined by a certain number of support vectors. Support vectors are a subset of learning data serving to define the limit between the two classes. When the support-vector machine cannot separate two classes, it solves this problem by mapping the input data in spaces of high-dimensional functions by means of a kernel function. In a high-dimensional space, it is possible to create a hyperplane enabling a linear separation (which corresponds to a curved surface in the lower input space). Consequently, the kernel function plays an important role in the support-vector machine. In practice, various kernel functions can be used, such as linear, polynomial, or Gaussian. A remarkable property of the support-vector machine is its learning capacity, which does not depend on the dimensionality of the characteristic space. This means that the support-vector machine can generalize when given numerous functionalities. Mukkamala and Sung (2003b) showed the many advantages of the support-vector machine compared to other techniques. Support-vector machines surpass neural networks in terms of upgradability, learning time, runtime and prediction accuracy. Mukkamala and Sung (2003a) also applied support-vector machines for the extraction of intrusion detection characteristics of KDD files. They empirically proved that the functionalities selected using the support-vector machine yielded similar results as the use of a full set of functionalities. This decrease in the number of functionalities reduces the computation efforts. Chen *et al.* (2005) also proved that support-vector machines surpassed neural networks.

1.3.5. *Clustering techniques*

Clustering techniques operate by organizing observed data in groups, depending on a given similarity or a distance measurement. Similarity can be measured by using the cosine formula, the binary weighted cosine formula proposed by Rawat (2005) or other formulas. The most commonly used procedure for clustering involves the selection of a representative point for each cluster. Then each new data point is classified as belonging to a given group depending on the proximity to the corresponding representative point. There are at least two approaches for the classification-based detection of anomalies. In the first approach, the anomaly detection model is formed using unlabeled data including both normal and attack traffic. In the second approach, the model is formed using only normal data and a normal activity profile is created. The idea underlying the first approach is that abnormal or attack data represent a small percentage of the total data. If this hypothesis is verified, anomalies and attacks can be detected depending on cluster size: large clusters correspond to normal data and the other data points to attacks. Liao and Vemuri (2002) used the *K-nearest neighbor* (K-nn) approach, based on the Euclidian distance, to define the belonging of data points to a given cluster. The Minnesota intrusion detection system is a network-based anomaly detection approach that uses data exploration and clustering techniques (Levent *et al.* 2004).

Leung and Leckie (2005) proposed an unsupervised anomaly detection approach for intrusion detection on a network. The proposed algorithm, known as "fpMAFIA", is a clustering algorithm based on density and on grid for large data sets. The major advantage of this algorithm is that it can produce arbitrary forms and cover over 95% of the set of data with appropriate values of parameters. The authors proved that the algorithm evolves linearly with respect to the number of registrations in the set of data. They evaluated the accuracy of the newly proposed algorithm and proved that it enables reaching a reasonable detection rate.

1.3.6. *Hybrid techniques*

Many researchers suggested that the monitoring capacity of current IDS systems could be improved by adopting a hybrid approach including detection techniques of both anomalies and signatures (Lunt *et al.* 1992; Anderson *et al.* 1995; Fortuna *et al.* 2002; Hwang *et al.* 2007). Sabhnani and Serpen (2003) proved that no single classification technique enables the detection of all the attack classes at an acceptable false alarm rate and with a good detection accuracy. The authors used various techniques to classify the intrusions by means of a KDD 1998 dataset. Many researchers proved that the hybrid or set-based classification technique can improve detection accuracy (Mukkamala *et al.* 2005; Chen *et al.* 2005; Aslahi-Shahri *et al.* 2016; Hamamoto *et al.* 2018; Hajimirzaei and Navimipour 2019; Sai Satyanarayana Reddy *et al.* 2019). A hybrid approach involves the integration of various learning or decision-making models. Each learning model operates differently and uses a different set of functionalities. The integration of various learning models yields better results than the individual learning or decision-making models and reduces their individual limitations. A significant advantage of the combination of redundant and complementary classification techniques is that it increases robustness and accuracy in most applications.

Various methods combining various classification techniques were proposed in the literature (Menahem *et al.* 2009; Witten *et al.* 2016). Ensemble methods have a common objective: to build a combination of certain models, instead of using a single model to improve the results. Mukkamala and its collaborators (2005) proved that the use of ensemble classifiers led to the best possible accuracy for each category of attack models. Chebrolu *et al.* (2005) used the Classification And Regression Trees-Bayesian network (CART-BN) approach for intrusion detection. Zainal *et al.* (2009) proposed the hybridization of linear genetic programming of the adaptive neural fuzzy inference system and of random forests for intrusion detection. They proved empirically that by assigning appropriate weights to the classifiers in a hybrid approach, the accuracy of detection of all the classes of network traffic is improved compared to an individual classifier. Menahem *et al.* (2009) used various classifiers and tried to take advantage of their strengths. Hwang *et al.* (2007) proposed a three-level hybrid approach to detect intrusions. The first level of the system is a signature-based approach in order to filter the known attacks using the black list concept. The second level of the system is an anomaly detector that uses the white list concept to distinguish between the normal traffic and the attack traffic surpassed by the first level. The third level of the system uses support vectors machines in order to classify the unknown attack traffic. The success of a hybrid method depends on many factors, notably the size of the learning sample, the choice of a basic classifier, the exact manner in which the forming set is modified, the choice of combination method and finally the data distribution and the potential capacity of the basic classifier chosen for solving the problem (Rokach 2010).

1.4. AI misuse

AI is a double use domain. AI systems and the manner in which they are designed can serve both civilian and military purposes, and in a broader sense, beneficial or harmful purposes. Given that certain tasks requiring intelligence are benign while others are not, AI is double edged in the same way that human intelligence is. Researchers in the field of AI cannot avoid producing systems that can serve harmful purposes. For example, the difference between the capacities of an autonomous drone used for delivering parcels and the capacities of an autonomous drone used for delivering explosives is not necessarily too wide. Moreover, fundamental research aiming to improve our comprehension of AI, its capacities and its control seem to be inherently double edged.

AI and machine learning have an increasingly important impact on the security of citizens, organizations and states. Misuse of AI will impact the way in which we build and manage our digital infrastructure, as well as the design and distribution of AI systems, therefore it will probably require an institutional policy. It is worth noting here that the threats caused by AI misuse have been highlighted in heavily publicized contexts (for example, during a Congress hearing (Moore and Anderson 2012), a workshop organized by the White House and a report of the US Department for Homeland Security).

The increasing use of AI for the development of cyberattack techniques and the absence of development of adequate defenses has three major consequences.

1.4.1. *Extension of existing threats*

For many known attacks, the progress of AI is expected to enlarge the set of players capable of conducting the attack, their attack speed and the set of possible targets. This is a consequence of the efficiency, upgradability and ease of dissemination of AI systems. In particular, the dissemination of intelligent and efficient systems can increase the number of players who can afford specific attacks. If the reliable intelligent systems are also evolutionary (upgradable), then even the players who already have the required resources to conduct these attacks may acquire the capacity to execute them at a much faster pace.

An example of a threat that is susceptible to develop in this manner is the phishing attack threat. These attacks use personalized messages to obtain sensitive information or money from their victims. The attacker often introduces himself as one of the friends, colleagues or professional contacts of the target. The most advanced phishing attacks require significant qualified manpower, as the attacker must identify the high value targets, research their social and professional networks, and then generate messages that are acceptable to the target.

1.4.2. *Introduction of new threats*

AI progress will enable new varieties of attacks. These attacks may use AI systems to conduct certain tasks more successfully than any human being.

Due to their unlimited capacities, in contrast with those of humans, intelligent systems could enable players to conduct attacks that would otherwise be impossible. For example, most persons are not able to efficiently imitate the voice of other persons. Consequently, the creation of audio files resembling recordings of human speech becomes essential in these cases. Nevertheless, significant progress has been recently achieved in the development of speech synthesis systems, which learn to imitate human voice. Such systems would in turn enable new methods for spreading disinformation and imitating others.

Moreover, AI systems could be used to control certain aspects of malware behavior that would be impossible to control manually. For example, a virus designed to modify the behavior of ventilated computers, as in the case of the Stuxnet program, used to disrupt the Iranian nuclear program, cannot receive commands once these computers are infected. Limited communication problems also occur under water and in the presence of signal jammers.

1.4.3. *Modification of the typical threat character*

Properties of AI such as efficiency, upgradability and capacities surpassing those of humans may enable very relevant attacks. Attackers are often faced with a compromise between the frequency, the extent of their attacks and their efficiency. For example, spear phishing is more effective than classical phishing, which does not involve adapting messages to individuals, but it is relatively costly and cannot be conducted en mass. More generic phishing attacks are profitable despite their very low success rates, simply because of their extent. If the frequency and upgradability of certain attacks, including spear phishing, are improved, AI systems can mitigate these compromises. Moreover, properties such as efficiency and upgradability, particularly in the context of target identification and analysis, lead also to finely targeted attacks. The attackers are often interested in adapting their attacks to the characteristics of their targets, aiming at targets with certain properties, such as significant assets or an association with certain political groups. Nevertheless, the attackers must often find a balance between efficiency, the upgradability of their attacks and target precision. A further example could be the use of drone swarms that deploy facial recognition technology to kill specific individuals in a crowd, instead of less targeted forms of violence.

Cyberattacks are increasingly alarming in terms of complexity and quantity, a consequence of the lack of awareness and understanding of the actual needs. This lack of support explains the insufficient dynamism, attention and willingness to commit funds and resources for cybersecurity in many organizations. In order to limit the impact of cyberattacks, the following recommendations are suggested (Brundage *et al.* 2018):

– decision-makers should closely cooperate with technical researchers to study, prevent and limit the potential misuse of AI;

– researchers and engineers in the AI field should seriously consider the double-edged nature of their work, by allowing considerations linked to abusive use to influence the research priorities and norms and by proactively addressing concerned players when harmful applications are predictable;

– public authorities should actively try to broaden the range of stakeholders and experts in the field that are involved in the discussions related to these challenges.

1.5. Conclusion

AI is a broad domain to be explored by cybersecurity researchers and experts. As the capacity of intelligent systems increases, they will first reach and then surpass human capacities in many fields. In cybersecurity, AI can be used to strengthen the defenses of computer infrastructure. It is worth noting that, as AI covers fields considered reserved to humans, the security threats will increase in variety, difference and intelligence compared to actually existing techniques. Defense against these threats is very difficult, as cybersecurity experts themselves can be targeted by spear phishing attacks. Consequently, preparing for potential misuses of AI associated with this transition is an important task. The use of intelligent techniques aims to identify real-time attacks, with little or no human interaction, and to stop them before they cause damages. In conclusion, AI can be considered as a powerful tool in solving cybersecurity problems.

1.6. References

Agarwal, R. and Joshi, M.V. (2000). A new framework for learning classifier models in data mining [Online]. Available at: https://pdfs.semanticscholar.org/db6e/1d67f7912efa65f948 07dc81b24dea2de158.pdf [Accessed January 2019].

Ahlan, A.R., Lubis, M., and Lubis, A.R. (2015). Information security awareness at the knowledge-based institution: Its antecedents and measures. *Procedia Computer Science (PCS)*. 72(2015), 361–373.

Amodei, D., Olah, C., Steinhardt, J., Christiano, P., Schulman, J., and Mané. (2016). Concrete problems in AI safety [Online]. Cornell University. Available at: https://arxiv.org/abs/ 1606.06565.

Anderson, D., Frivold, T., and Valdes, A. (1995). Next-generation intrusion detection expert system (NIDES). Report, US Department of the Navy, Space and Naval Warfare Systems Command, San Diego.

Aslahi-Shahri, B.M., Rahmani, R., Chizari, M., Maralani, A., Eslami, M., Golkar, M.J., and Ebrahimi, A. (2016). A hybrid method consisting of GA and SVM for intrusion detection system. *Neural Computing and Applications*, 27(6), 1669–1676.

Bace, R.G. (2000). *Intrusion Detection*. Sams Publishing, Indianapolis.

Balan, E.V., Priyan, M.K., Gokulnath, C., and Devi, G.U. (2015). Fuzzy based intrusion detection systems in MANET. *Procedia Computer Science*, 50, 109–114.

Barth, C.J. and Mitchell, J.C. (2008). Robust defenses for cross-site request forgery. *Proceedings of 15th ACM Conference*. CCS, Alexandria.

Biggio, B., Nelson, B., and Laskov, P. (2012). Poisoning attacks against support vector machines. *29th International Conference on Machine Learning*. ICML, Edinburgh, 1467–1474.

Capgemini Research Institute (2019). Reinventing cybersecurity with artificial intelligence: The new frontier in digital security [Online]. Available at: https://www.capgemini.com/ wp-content/uploads/2019/07/AI-in-Cybersecurity_Report_2019 0711_V06.pdf.

Chebrolu, S., Abraham, A., and Thomas. (2005). Feature deduction and ensemble design of intrusion detection systems. *Computers & Security*, 24(4), 295–307.

Chen, W.-H., Hsu, S.-H., and Shen, H.-P. (2005). Application of SVM and ANN for intrusion detection. *Computers & Operations Research*, 32(10), 2617–2634.

Cova, M., Balzarotti, D., Felmetsger, V., and Vigna, G. (2007). Swaddler: An approach for the anomaly-based detection of state violations in web applications. *Proceedings of the 10th International Symposium on Recent Advances in Intrusion Detection.* RAID, Gold Coast.

Cova, M., Kruegel, C., and Vigna, G. (2010). Detection and analysis of drive-by-download attacks and malicious JavaScript code. *Proceedings of the 19th International Conference on the World Wide Web.* WWW, Raleigh.

Crockford, D. (2015). Json [Online]. Available at: https://github.com/douglascrockford/JSON-js/blob/master/README [Accessed March 2018].

Cunningham, R. and Lippmann, R. (2000). Detecting computer attackers: Recognizing patterns of malicious stealthy behavior. Presentation, CERIAS, Anderlecht.

Ertoz, L., Eilertson, E., Lazarevic, A., Tan, P.N., Kumar, V., Srivastava, J., & Dokas, P. (2004). Minds-Minnesota intrusion detection system. *Next Generation Data Mining*, August, 199–218.

Fortuna, C., Fortuna, B., and Mohorčič, M. (2002). Anomaly detection in computer networks using linear SVMs [Online]. Available at: http://ailab.ijs.si/dunja/SiKDD2007/Papers/Fortuna_Anomaly.pdf.

Hajimirzaei, B. and Navimipour, N.J. (2019). Intrusion detection for cloud computing using neural networks and artificial bee colony optimization algorithm. *ICT Express*, 5(1), 56–59.

Hamamoto, A.H., Carvalho, L.F., Sampaio, L.D.H., Abrão, T., & Proença Jr, M.L. (2018). Network anomaly detection system using genetic algorithm and fuzzy logic. *Expert Systems with Applications*, 92, 390–402.

Han, X., Xu, L., Ren, M., and Gu, W. (2015). A Naive Bayesian network intrusion detection algorithm based on principal component analysis. *7th International Conference on Information Technology in Medicine and Education*. IEEE, Huangshan.

Heckerman, D. (2008). A tutorial on learning with Bayesian networks. *Innovations in Bayesian Networks*, Holmes, D.E. and Jain, L.C. (eds). Springer, Berlin, 33–82.

Hoque, M.S. *et al.* (2012). An implementation of intrusion detection system using genetic algorithm. *International Journal of Network Security & Its Applications (IJNSA)*. AIRCC publisher, 4(2), 109–120.

Hu, J., Yu, X., Qiu, D., and Chen, H.H. (2009). A simple and efficient hidden Markov model scheme for host-based anomaly intrusion detection. *IEEE Network*, 23(1), 42–47.

Hwang, T.S., Lee, T.-J., and Lee, Y.-J. (2007). A three-tier IDS via data mining approach. *Proceedings of the 3rd Annual ACM Workshop on Mining Network Data*. ACM, San Diego.

Idris, N.B. and Shanmugam, B. (2005). Artificial intelligence techniques applied to intrusion detection. *Annual IEEE India Conference (Indicon)*. IEEE, Chennai.

Ingham, K., Somayaji, A., Burge, J., and Forrest, S. (2007). Learning DFA representations of HTTP for protecting web applications. *Journal of Computer Networks*, 51(5), 1239–1255.

Kalaivani, S., Vikram, A., and Gopinath, G. (2019). An effective swarm optimization based intrusion detection classifier system for cloud computing. *5th International Conference on Advanced Computing & Communication Systems (ICACCS)*. IEEE, Coimbatore.

Kang, M.-J. and Kang, J.-W. (2016). Intrusion detection system using deep neural network for in-vehicle network security. *PLOS ONE*, 11(6), 1–17.

Khidzir, N.Z., Daud, K.A.M., Ismail, A.R., Ghani, M.S.A.A., and Ibrahim, M.A.H. (2018). Information Security Requirement: The Relationship Between Cybersecurity Risk Confidentiality, Integrity and Availability in Digital Social Media. *Regional Conference on Science, Technology and Social Sciences (RCSTSS 2016)*. 4–6 December 2016, Penang, Malaysia.

Kim, J., Kim, J., Thu, H.L.T., and Kim, H. (2016). Long short term memory recurrent neural network classifier for intrusion detection. *International Conference on Platform Technology and Service (PlatCon)*. IEEE, Jeju.

Kirdaa, E., Jovanovicb, N., Kruegelc, C., and Vigna, G. (2009). Client-side cross-site scripting protection. *Computers & Security*, 28(7), 592–604.

Kruegel, C., Mutz, D., Robertson, W., and Valeur, F. (2003). Bayesian event classification for intrusion detection. *Proceedings of the 19th Annual Computer Security Applications Conference*. IEEE, Las Vegas.

Kudłacik, P., Porwik, P., and Wesołowski, T. (2016). Fuzzy approach for intrusion detection based on user's commands. *Soft Computing*, 20(7), 2705–2719.

Kumar, S., Krishna, C.R., and Solanki, A.K. (2018). A technique to resolve data integrity and confidentiality issues in a wireless sensor network. *8th International Conference on Cloud Computing, Data Science & Engineering (Confluence)*. IEEE, Noida.

Landwehr, C. (2008). Cybersecurity and artificial intelligence: From fixing the plumbing to smart water. *IEEE, Security and Privacy*, 6(5), 3–4.

Leung, K. and Leckie, C. (2005). Unsupervised anomaly detection in network intrusion detection using clusters. *Proceedings of the 28th Australasian Conference on Computer Science*. Australian Computer Society Inc., Darlinghurst, 333–342.

Li, L., De-Zhang, Y. and Chen, F.-S. (2010). A novel rule-based Intrusion Detection System using data mining. *3rd International Conference on Computer Science and Information Technology*. IEEE, Chengdu.

Liang, J. *et al.* (2019). A filter model for intrusion detection system in vehicle ad hoc networks: A hidden Markov methodology. *Knowledge-Based Systems*, 163, 611–623.

Liao, Y. and Vemuri, V.R. (2002). Use of k-nearest neighbor classifier for intrusion detection. *Computers & Security*, 21(5), 439–448.

Lippmann, R.P. and Cunningham, R.K. (2000). Improving intrusion detection performance using keyword selection and neural networks. *Computer Networks*, 34(4), 597–603.

Lunt, T. (1993). Detecting intruders in computer systems. *Proceedings of the 1993 Conference on Auditing and Computer Technology*. Baltimore Convention Center, Baltimore.

Lunt, T.F. (1990). Real-time intrusion detection expert system. Computer Science Lab., SRI International, Technical Report.

Mahoney, M.V. and Chan, P.K. (2001). PHAD: Packet header anomaly detection for identifying hostile network traffic [Online]. Available at: https://pdfs.semanticscholar.org/1505/f3658f5af7dff88e88d6a2b381de12e03036.pdf.

Mahoney, M.V. and Chan, P.K. (2002a). Learning models of network traffic for detecting novel attacks. Technical Report, Florida Institute of Technology, Melbourne.

Mahoney, M.V. and Chan, P.K. (2002b). Learning nonstationary models of normal network traffic for detecting novel attacks. *Proceedings of the 8th ACM SIGKDD International Conference on Knowledge Discovery and Data Mining*. ACM, Edmonton.

Menahem, E., Shabtai, A., Rokach, L. and Elovici, Y. (2009). Improving malware detection by applying multi-inducer ensemble. *Computational Statistics & Data Analysis*, 53(4), 1483–1494.

Miles, B., Shahar, A., Jack, C., Helen, T., Peter, E., Ben, G., Allan, D., Paul, S., Thomas, Z., Bobby, F., Hyrum, A., Heather, R., Gregory, C.A., Jacob, S., Carrick, F., Seán, Ó. h., Simon, B., Haydn, B., Sebastian, F., Clare, L., Rebecca, C., Owain, E., Michael, P., Joanna, B., Roman, Y. and Dario, A. (2018). The malicious use of artificial intelligence: Forecasting, prevention, and mitigation [Online]. Available at: https://arxiv.org/ftp/arxiv/papers/1802/1802.07228.pdf.

Mishra, A., Agrawal, A. and Ranjan, R. (2011). Artificial intelligent firewall. *Proceedings of the International Conference on Advances in Computing and Artificial Intelligence*. ACM, Rajpura/Punjab.

Moon, D., Im, H., Kim, I. and Park, J. H. (2017). DTB-IDS: An intrusion detection system based on decision tree using behavior analysis for preventing APT attacks. *The Journal of Supercomputing*, 73(7), 2881–2895.

Moore, T. and Anderson, R. (2012). *Internet Security. The Oxford Handbook of the Digital Economy*. Oxford University Press, Oxford.

Mukkamala, S. and Sung, A.H. (2003a). Artificial intelligent techniques for intrusion detection. *International Conference on Systems, Man and Cybernetics*. IEEE, Washington.

Mukkamala, S. and Sung, A.H. (2003b). A comparative study of techniques for intrusion detection. *Proceedings of the 15th IEEE International Conference on Tools with Artificial Intelligence (ICTAI'03)*. IEEE, Washington.

Mukkamala, S., Sung, A.H., and Abraham, A. (2005). Intrusion detection using an ensemble of intelligent paradigms. *Journal of Network and Computer Applications*, 28, 167–182.

Mutz, D., Robertson, W., Vigna, G., and Kemmerer, R. (2007). Exploiting execution context for the detection of anomalous system calls. *Proceedings of the International Symposium on Recent Advances in Intrusion Detection*. RAID, Gold Coast.

Novikov, D., Yampolskiy, R.V., and Reznik, L. (2006). Artificial intelligence approaches for intrusion detection. *IEEE Long Island Systems, Applications and Technology Conference*. IEEE, Long Island.

Peltier, T.R. (2010). *Information Security Risk Analysis*. CRC Press, Boca Raton.

Peng, K., Leung, V., Zheng, L., Wang, S., Huang, C., and Lin, T. (2018). Intrusion detection system based on decision tree over big data in fog environment [Online]. Available at: https://www.hindawi.com/journals/wcmc/2018/ 4680867/.

Ponkarthika, M. and Saraswathy, V.R. (2018). Network intrusion detection using deep neural networks. *Asian Journal of Applied Sciences*, 2(2), 665–673.

Quamar, N., Weiqing, S., Ahmad, Y.J., and Mansoor, A. (2016). A deep learning approach for network intrusion detection system. *Proceedings of the 9th EAI International Conference on Bio-inspired Information and Communications Technologies (formerly BIONETICS)*. ICST publisher, December 3–5, 2015, New York, USA, 21–26.

Rai, K., Devi, M.S., and Guleria, A. (2016). Decision tree based algorithm for intrusion detection. *International Journal of Advanced Networking and Applications*, 7(4), 2828.

Rawat, S. (2005). Efficient data mining algorithms for intrusion detection. *Proceedings of the 4th Conference on Engineering of Intelligent Systems (EIS 2004)*. EIS, Madeira.

Robertson, W., Maggi, F., Kruegel, C., and Vigna, G. (2010). Effective anomaly detection with scarce training data. *Proceedings of the Network and Distributed System Security Symposium*, NDSS, San Diego.

Roesch, M. (1999). Snort: Lightweight intrusion detection for networks. *Lisa*, 99(1), 229–238.

Rokach, L. (2010). Ensemble-based classifiers. *Artificial Intelligence Review*, 33(1/2), 1–39.

Sabhnani, M. and Serpen, G. (2003). Application of machine learning algorithms to KDD intrusion detection dataset within misuse detection context. *International Conference on Machine Learning; Models, Technologies and Applications*. MLMTA, Las Vegas.

Sahu, S. and Mehtre, B.M. (2015). Network intrusion detection system using J48 Decision Tree. *International Conference on Advances in Computing, Communications and Informatics (ICACCI)*. IEEE, Kochi.

Sai Satyanarayana Reddy, S., Chatterjee, P., and Mamatha, C. (2019). Intrusion detection in wireless network using fuzzy logic implemented with genetic algorithm. In *Computing and Network Sustainability*, Peng, S.-L, Dey, N., and Bundele, M. (eds). Springer, Berlin, 425–432.

Scharre, P. (2015). *Counter-swarm: A guide to defeating robotic swarms* [Online]. Available at: https://warontherocks.com/2015/03/counter-swarm-a-guide-todefeating-robotic-swarms/.

Schneier, B. (2008). The psychology of security. *International Conference on Cryptology in Africa*. AFRICACRYPT, Casablanca.

Shanmugavadivu, R. and Nagarajan, N. (2011). Network intrusion detection system using fuzzy logic. *Indian Journal of Computer Science and Engineering*, 2(1), 101–111.

Szegedy, C., Zaremba, W., Sutskever, I., Bruna, J., Erhan, D., Goodfellow, I. and Fergus, R. (2013). Intriguing properties of neural networks [Online]. Available at: https://arxiv.org/abs/1312.6199.

Tekerek, A. and Bay, O.F. (2019). Design and implementation of an artificial intelligence-based web application firewall model. *Neural Network World*, 189, 206.

Teng, H.S. and Chen, K. (1990). Adaptive real-time anomaly detection using inductively generated sequential patterns. *Proceedings of the 1990 IEEE Computer Society Symposium on Research in Security and Privacy*. IEEE, Oakland.

Turner, C., Jeremiah, R., Richards, D., and Joseph, A. (2016). A rule status monitoring algorithm for rule-based intrusion detection and prevention systems. *Procedia Computer Science*, 95, 361–368.

Valentín, K. and Malý, M. (2014). Network firewall using artificial neural networks. *Computing and Informatics*, 32(6), 1312–1327.

Vapnik, V. (1998). *Statistical Learning Theory*. John Wiley and Sons, Hoboken.

Veiga, A.P. (2018). Applications of artificial intelligence to network security [Online]. Available at: https://arxiv.org/abs/1803.09992.

Vinayakumar, R., Soman, K.P. and Poornachandran, P. (2017). Applying convolutional neural network for network intrusion detection. *6th International Conference on Advances in Computing, Communications and Informatics (ICACCI)*. Manipal University, Karnataka.

Witten, I.H. and Frank, E. (2016). *Data Mining: Practical Machine Learning Tools and Techniques*. Morgan Kaufmann, Burlington.

Yang, Y., McLaughlin, K., Littler, T., Sezer, S. and Wang, H.F. (2013). Rule-based intrusion detection system for SCADA networks. *2nd IET Renewable Power Generation Conference (RPG 2013)*. RPG, Beijing.

Zainal, A., Maarof, M.A. and Shamsuddin, S.M. (2009). Ensemble classifiers for network intrusion detection system. *Journal of Information Assurance and Security*, 4(3), 217–225.

Zegeye, W.K., Moazzami, F. and Dean, R. (2018). Hidden Markov Model (HMM) based Intrusion Detection System (IDS). *International Telemetering Conference Proceedings*, 5–8 November 2018, Glendale, Arizona.

2

An Intelligent Control Plane for Security Services Deployment in SDN-based Networks

Maïssa MBAYE[1], Omessaad HAMDI[2] and Francine KRIEF[3]

[1] *Gaston Berger University, Saint-Louis, Senegal*

[2] *IMT Atlantique, Rennes, France*

[3] *ENSEIRB-MATMECA, Bordeaux, France*

2.1. Introduction

The software-defined networking (SDN) approach involves the management of a network infrastructure by software applications (Fortes 2013; ONF 2015). These applications may be hosted by one or several servers and enable the control of the (physical or virtual) network components of the infrastructure. This new model of network architecture, relying on the separation between the control plane and the data plane, is increasingly drawing the attention of major players in networking and telecommunications (Open Networking Foundation [ONF], IETF, ITU-T, ETSI) (Kreutz *et al.* 2015).

The data plane corresponds to network equipment in charge of optimized (at best) data transfer to the destination node. In this plane, the main task involves using a forwarding/flow table for correct data retransmission. In the case of SDN/OpenFlow-based networks, this table contains the flow transfer rules that match the fields values of common headers (Ethernet, IPv4/v6, MPLS, TCP/UDP, etc.) and deduces from them the action to be taken (retransmit, delete, modify header, etc.). From 2015 onward, the main manufacturers (IBM, Hewlett-Packard, Huawei,

Intelligent Network Management and Control,
coordinated by Badr BENMAMMAR.

Juniper, NEC, etc.) launched SDN/OpenFlow compatible network equipment on the market (Kreutz *et al.* 2015).

The control plane represents the entire logic enabling the management of the forwarding table content. It has an overall view of the network and includes the protocols enabling the data plane programming depending on the needs of the applications deployed in the network (routing, access control, quality of service [QoS], load balancing, traffic engineering, security, etc.). This control plane is implemented by a software platform that is (logically) centralized in a cloud and is known as controller. This plane offers innovative functionalities, such as network virtualization.

The main benefits of SDN-based networks and network virtualization are that the new generation networks are more flexible, agile and adaptable and highly automated. Indeed, the transition from IPv4 to IPv6 shows that because equipment manufacturers are responsible for the deployment of new protocols and network applications in their products, the time required to launch a product on the market is longer. With this new approach, it is possible for each structure to have new network applications and protocols programmed on the equipment eliminating the need to wait for a newer version of the equipment. Nevertheless, this possibility raises a certain number of challenges for security and scalability.

Security challenges mainly take two forms: SDN-based network security and deployment of security services. SDN-based network security efforts are essentially focused on securing the controller, which is a centralized decision point. Indeed, if the controller is compromised, out of service or disconnected from the data plane, network behavior consistency is no longer guaranteed. The controller may be the target of different kinds of attacks, such as distributed denial of service (DDoS) attacks, unauthorized access and injection in security policies (Shu *et al.* 2016). The solutions proposed cover a wider range, from firewalls (Wang *et al.* 2013) to intrusion detection/prevention systems (IDS/IPS) (Gowtham *et al.* 2018), in order to counteract the threats targeting SDN-based network security.

In SDN-based networks, the deployment of security applications (for example, IDS/IPS) aims to offer security as an SDN-based network service to the customers. The two major challenges of security services deployment are as follows: the dynamic character of SDN-based network control and the large amount of data to be processed by these solutions. This involves the deployment of security services that will offer acceptable response and processing times.

In classical networks, machine learning (ML) and artificial intelligence (AI) have generally proved their effectiveness for security (Das and Nene 2017). Nowadays,

AI is increasingly used for solving any type of problem, partly due to the amount of data generated through the Internet. SDN-based network security is also in line with this trend, and several works already approach the SDN-based network security problem, providing solutions that rely on AI tools (Abubakar and Pranggono 2018; Dey *et al.* 2018; Xie *et al.* 2019). The ultimate objective would be to have intelligent SDN-based networks with self-protection and self-optimization capabilities.

This chapter aims to approach AI-based intelligent control techniques that enable an intelligent management of security deployment.

As a first step, section 2.2 offers a presentation of SDN-based networks, followed by an approach to the security problem and the description of an architecture for the deployment of security services for SDN-based networks in section 2.3. Section 2.4 presents the various initiatives aimed at making SDN-based networks "intelligent". Section 2.5 focuses on the contribution of AI to network security, while section 2.6 deals with security in SDN-based networks. We describe our proposal for the deployment of an intelligent service for intrusion prevention in section 2.7. Section 2.8 deals with the main challenges to be addressed in view of the wide use of AI tools for SDN-based network security. Finally, section 2.9 concludes this chapter.

2.2. Software-defined networking

The SDN-based network concept marks a significant development of traditional network architectures. This section presents the main concepts and architectural elements.

2.2.1. *General architecture*

SDN architecture is mainly composed of three planes (Figure 2.1): the application plane, data plane, and control plane. Separation between the control plane and the data plane is at the core of this architecture (ONF 2015).

The *application plane* contains both the SDN applications and the network applications. The SDN controller uses an application programming interface (API), known as northbound API, which interacts with the application plane. This plane contains applications enabling the implementation of purely network functionalities, such as QoS and routing. Other SDN applications whose role is to control the SDN-based network logic can also be found here (Bannour *et al.* 2018).

The *data plane* is composed of physical or virtual equipment for interconnection with a main task of data forwarding. In fact, in traditional networks, the network equipment contains the control plane and the data plane. With this new approach, the control plane is externalized in order to enhance network equipment's efficiency. These network elements (mainly switches) contain the forwarding tables and can be remotely controlled via APIs.

The *control plane* is implemented at the SDN controller level, generally on a physical server or in the cloud. The controller manages the entire "intelligent" logic of the SDN-based network by programming (on-the-fly manipulation) the content of the forwarding tables at the data plane level. The SDN controller manages the infrastructure in its entirety and is able to acquire real-time information on the state and activity of the (physical or virtual) network equipment it controls.

In a routing protocols context, the control plane would correspond to the component whose task it is to find the best paths for the router and set a ready to use routing table for it. It is logically centralized but can be physically distributed between several elements (Bannour *et al.* 2018). When controllers are physically distributed, they can communicate using East and Westbound API.

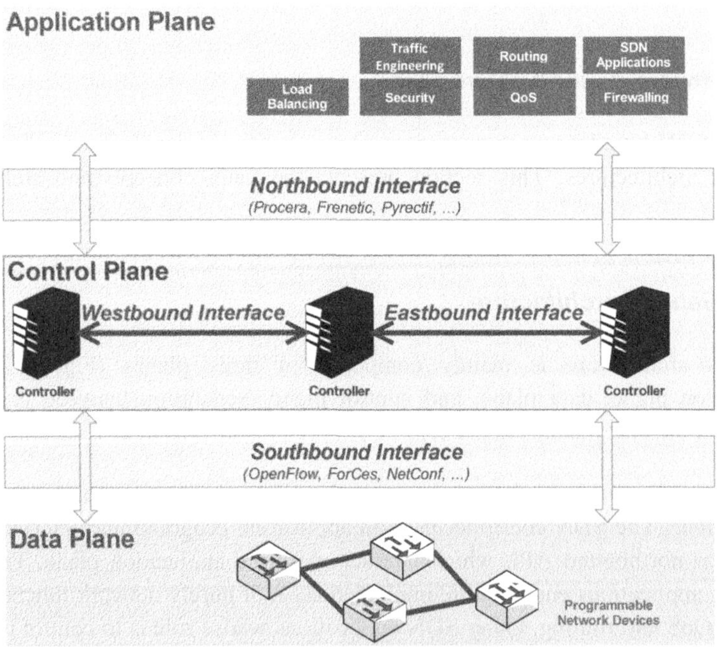

Figure 2.1. *Simplified SDN architecture (Zhang* et al. *2017)*

The control plane and the data plane communicate via a class of protocols known as a Southbound interface. In 2019, the most advanced and commercially available Southbound interface for SDN-based networks was OpenFlow[1]. There are however other alternatives such as ForCES, SNMP and NetConf of IETF. Furthermore, other approaches to network equipment programming exist, such as P4 (Bosshart *et al.* 2014; Cordeiro *et al.* 2017), which are currently under development.

OpenFlow was originally a Stanford University project. Today, it is a standard network protocol, published by the ONF, which serves as a link between the control plane and the data plane. This protocol is composed of instructions (rules) that enable the programming of forwarding tables of network equipment, referred to here as flow tables. The instructions define an action on the traffic (packet transmission, packet rejection, etc.). Various versions of the OpenFlow protocol have been introduced to add more flexibility and reliability including several flow tables, improved correspondence/action capacities, optical ports, group tables, etc.[2]

Furthermore, many OpenFlow controllers are available, such as POX (Kaur *et al.* 2014), Beacon (Erickson 2013) and OpenDayLight[3].

2.2.2. *Logical distribution of SDN control*

The physical centralization of the control plane in a single programmable software component known as a "controller" raises a number of problems, particularly in terms of scalability, availability and reliability (Nkosi *et al.* 2016; Karakus and Durresi 2017). Hence, the higher the number of switches, the higher the load of the SDN controller, and therefore its overload in terms of bandwidth, processing power and memory capacity. The communication delays between the SDN controller and the switches increase also with the network size and extent, which consequently influences the data flow latency (Bannour *et al.* 2018). The scalability problem can be solved by extending the responsibilities of the data plane in order to lighten the controller load, but this requires a modification of the switches design (Rebecchi *et al.* 2017).

The control plane should therefore be designed as a distributed system in which several SDN controllers are in charge with the overall network management, while maintaining a logically centralized network view. Indeed, this solution provides a

1 Available at: www.opennetworking.org/software-defined-standards/specifications/.

2 For further details on the various versions of the OpenFlow specification, please refer to: www.opennetworking.org/software-defined-standards/specifications/.

3 Available at: http://www.opendaylight.org/.

better scalability to the network control plane, while reducing the latencies of the control plane. Moreover, the use of several controllers improves reliability by eliminating the problem of the single point of failure.

Several works have focused on distributed control design. In Oktian *et al.* (2017), the various approaches to how a logically centralized view is provided to several instances of distributed controllers depending on the choice of design implemented have been discussed. They have thus identified two distributed architectures: flat architecture and hierarchical architecture (Figure 2.2). In flat architecture, each controller manages a subnetwork/field of the global network. In hierarchical architecture, local controllers manage the needs of local applications, while the main controller, generally known as the "root", deals with the needs of applications requiring a global network view.

In Bannour *et al.* (2018), the authors classified control architectures depending on how knowledge is disseminated between the controller instances. Two architectures are mentioned, the first centralized and the second distributed. This classification enabled them to compare the various SDN controller platforms in terms of upgradability, reliability and performance criteria. The authors prefer the hierarchical organization of the control plane for better upgradability and better performances. Each controller can thus have different responsibilities and make decisions based on a partial view of the network. The highest level acts as a centralized controller, with the problem of a single point of failure.

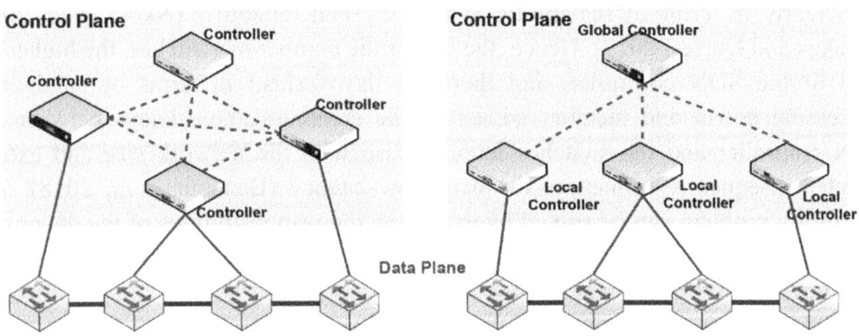

Figure 2.2. *Control plane distribution models*

Table 2.1 presents a classification of network information to be exchanged between distributed controllers in order to maintain the global view.

Category	Examples
State of the local network (static)	Reachability Topology Capacity QoS parameters
State of the local network (dynamic)	Flow tables Flow rules Real-time use of bandwidth Flow path
Controller event	State of the controller SDN application – specific information
Inventory	List of installed SDN applications List of connected switches

Table 2.1. *Classification of shared network information (Oktian* et al. *2017)*

The distribution of SDN control raises a number of challenges, among which the identification of the required number of controllers as well as their appropriate locations with respect to the expected performance and reliability objectives. Hence, the hierarchical organization of the control plane requires a relevant diagram of control distribution taking into account both the organization of SDN control plane and the physical location of SDN controllers (Bannour *et al.* 2018). Finally, to preserve the logically centralized view, the requirement to share knowledge between controllers may introduce new scalability problems. Indeed, the frequent propagation of state updates can make the network unavailable, which will increase the latency between controller and switches.

Another significant challenge relates to the fault tolerance of the distributed control architecture. Hence, controller coordination strategies should be implemented to reach agreements and also solve the problems related to the competition between updates and state consistency (Bannour *et al.* 2018). The state consistency of logically centralized SDN controllers is a major design challenge of SDN-based networks, which involves compromises between rule application and network performances. It is indeed very difficult to reach high consistency in an SDN environment subjected to network failures without compromising its availability and without adding to the complexity of the network state management. Recent approaches have introduced the concept of adaptive consistency, according to which the controllers can adjust their consistency level to reach the expected performance level, depending on specific metrics. This consistency should be taken into account when searching for the optimal location of controllers. Indeed, minimizing the distances between controllers is important for system performances; it facilitates the communication between controllers and improves the state consistency of the network. In Canini *et al.* (2014), the authors dealt with this problem and proposed a robust distributed SDN control plane, known as software transactional networking (STN).

The security of the distributed SDN-based network is a further crucial challenge. Decentralization of SDN control reduces the risk associated with a single point of failure and with attacks (DDoS, for example). However, the integrity of data flows between SDN controllers and switches is not always certain. A plausible scenario would be that of an attacker being able to corrupt a network by acting as an SDN controller.

Finally, it is also difficult to ensure the interoperability of distributed SDN controllers belonging to different SDN domains and using different controller technologies.

Several researches proposed the integration of automated and adaptive approaches in the distributed control plane in order to meet the previously mentioned challenges. Hence, in Ma *et al.* (2018), the authors use reinforced learning for the automation of the resource management and distribution in a distributed SDN-based network.

2.3. Security in SDN-based networks

Security is one of the factors hindering the deployment of SDN architectures, which are currently mainly centralized. The centralized aspect of the control plane has in fact many advantages, but it also raises a very important security problem, as all the control planes of the switches are located in a single point. Typically, the impact of the denial of service (DoS) attacks becomes very significant, as the control plane is no longer distributed, which represents a risk of compromising the entire network.

This section presents the various attack surfaces of SDN-based networks, before describing an architecture for the deployment of security services for SDN-based networks.

2.3.1. *Attack surfaces*

In terms of attack identification and response, SDN has two essential advantages, compared to traditional networks (Fortes 2013; Tang *et al.* 2016):

– the control plane enables an administrator to separate and block the attacks on any heterogeneous material (no need to individually reconfigure each component);

– instead of investing in an expensive intrusion detection system, SDN is able to distribute a task between the nodes; moreover, each node can become a firewall, a proxy, etc.

As for the drawbacks, SDN provides an opportunity to attackers when it exposes new interfaces, which means communication between the control plane and the data plane. By compromising the SDN controller, the entire network can be compromised. Consequently, when using SDN to provide intrusion detection systems (IDS) services, the security of the latter should be kept in mind. An impenetrable access control strategy for the SDN controller should therefore be designed and implemented.

Security of the application plane: the application plane includes various types of applications. Some of them play a very important role in the elaboration of flow rules. Attacking these applications may cause a malfunction of the SDN-based network. An attacker may inject a malicious code in the application or may illegally access the SDN-based network. Developing an access control model or a code verification mode is effective approach against this type of attacks (Klaedtke *et al.* 2014).

Security of controllers: the SDN controller is the most important element. An attacker conducting an attack on the controller may easily gain total control of the network. The vulnerabilities in the control plane may also lead to illegally controlling the controller. Moreover, an attacker may launch a flooding attack using the vulnerabilities of the switch (sending several packets of a compromised switch to a controller may disable it) or of the OpenFlow protocol (an attacker may send packets that do not correspond to the table and therefore the switch is going to send them all to the controller, which may trigger a DoS/DDoS). The controller is the main target of the DoS attacks.

Security of the data plane: the switch is the crucial part of the data plane. It transfers packets, reads the MAC address table and the ARP requests, etc. Several attacks are possible in the data plane, such as the DoS/DDoS. A further attack can be conducted via the size of the flow tables, which can be very significant and can therefore reduce the switching speed, especially when the Southbound interface is compromised.

Security of protocols: SDN protocols are mainly protocols of the Southbound and NorthBound interfaces. OpenFlow is typically a Southbound interface protocol. Despite its wide use, it still has vulnerabilities because there is no identification and access control for the communications between the switch and the SDN controller. Protocols can be secured by adding an access control mechanism to the network resources in the data plane from the control plane.

2.3.2. *Example of security services deployment in SDN-based networks: IPSec service*

Previous work by Coly and Mbaye (2019) proposes a framework for the deployment of security services in an SDN-based network. IPSec tunnels are taken as a security service example to illustrate the operation of this framework. Internet Protocol Security (IPsec) (Kent and Seo 2005) is a suite of protocols securing the IP layer of the TCP/IP model. This framework can be used to provide a virtual private network (VPN) or establish secure tunnels between two sites. This protocol uses the Internet Key Exchange (IKE) protocol for key negotiation and management.

The main outcomes of this proposal are as follows: design of an architecture for the deployment of security services in SDN-based networks, a new extension of OpenFlow protocol for the management of secured tunnels and, finally, integration of an IPSec-based tunnel mechanism in SDN-based networks as use case.

2.3.2.1. *General architecture*

The proposed architecture is presented in Figure 2.3. The networks (networks' costumers) are connected by an SDN core network. The core of the SDN-based network is composed of Border Gateway Switches (BGS) and Core Internal Switches (CIS). These switches communicate with the SDN controller through the OpenFlow Southbound interface. The SDN controller coordinates the deployment of security services in the SDN-based network. In the case of tunnels deployed between different end nodes, the security service is deployed only at BGS ingress and egress. In other cases, the service is deployed in all the CIS systems of the flow path.

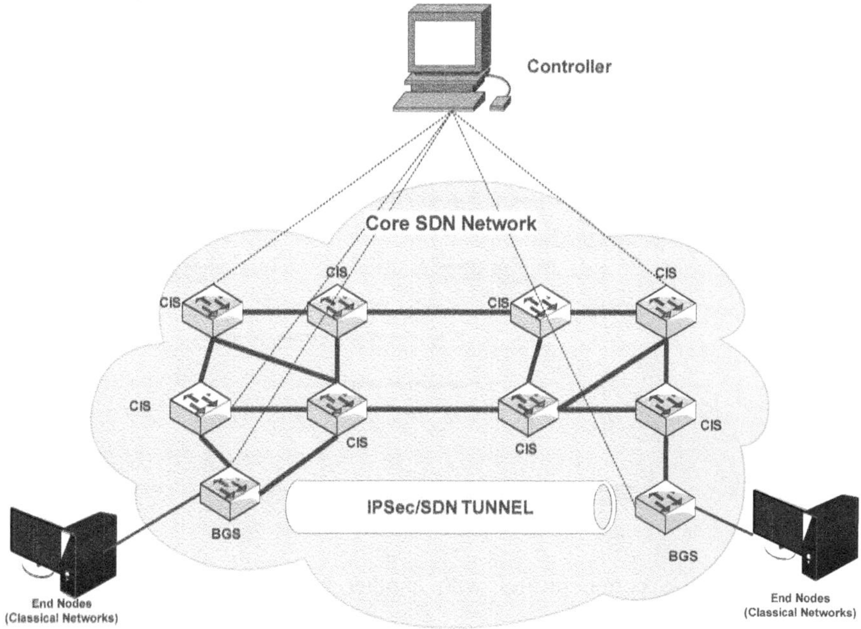

Figure 2.3. *Architecture of IPSec tunnel service deployment in an SDN-based network*

When a client subscribes to a service level agreement (SLA) comprising a security service, the SDN controller is set up according to this contract. The costumer's flows will henceforth drive the deployment of this service by sending a "SecTrans" message with the security policies to the (*ingress/egress*) BGS involved in this communication pathway. The remaining transaction is managed by the Southbound extension. As a use case illustrating the operation of this proposal, let us consider the deployment of an IPsec-based secured tunneling service, as SDN security service. For this purpose, we have extended the OpenFlow protocol for establishing an IPsec tunnel and defined an extended structure for the flow tables.

The SDN controller is in charge with the generation and transmission of IKE identification information. It is also responsible for the control and enforcement of IPsec SPD. Therefore, it has a centralized view of the network and security policies. The IKE component implemented in the network resource runs to create the IPSec security associations using these identification strategies and information. Figure 2.4 illustrates the tunnel deployment process.

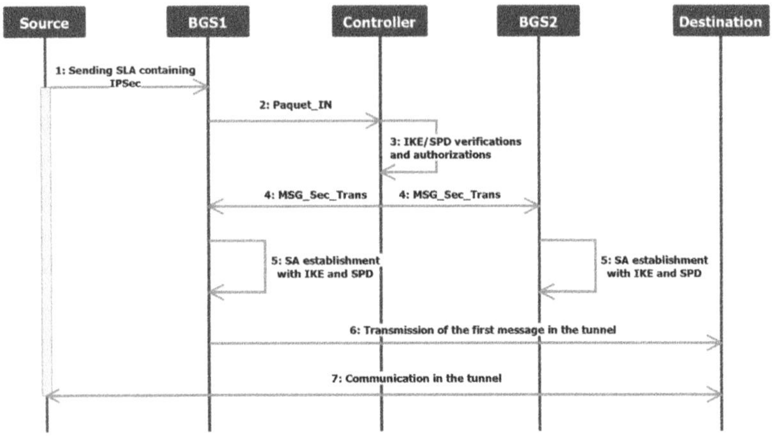

Figure 2.4. *Tunnel deployment process*

If an end node communicates with another one using an IPsec tunnel, the following procedure unfolds:

1) the BGS connected to the source sends a "Packet_In" message (an OpenFlow message) to the controller asking how these packets should be processed;

2) if there is a correspondence with the traffic of a service subscriber, the controller generates the IKE identification information and the SPD policies and then sends them to the involved (input and output) BGS in addition to the SecTrans message to allow transmission in the IPsec tunnel; the message adds/modifies a flow with "yes" in the IPsec field;

3) the BGS implements the IKE identification information and the SPD strategies to establish security associations before starting transmission.

Once the tunnel is set up, all the messages between these two termination points are transmitted through this tunnel.

2.3.2.2. *Deployment performance evaluation*

The proposal was evaluated using Mininet[4], OpenSwitch and Floodlight[5] as SDN controller. Table 2.2 summarizes the node configurations of the test bed.

4 Available at: http://www.mininet.org.

5 Floodlight OpenFlow Controller – Project Floodlight, Big switch network, available at: http://www.projectfloodlight.org/floodlight/.

Node	Operating system	(Software) Components	CPU	RAM
Controller	Debian 8.4	Floodlight master	(4) @ 3.2 GHz	4 GB
Machine A	Debian 8.4	Mininet 2.2.2, Racoon, ipsec-tools	(4) @ 3.2 GHz	4 GB
Machine B	Ubuntu 18.04	Mininet 2.2.2, Racoon, ipsec-tools	(4) @ 2.4 GHz	4 GB
End node A	Ubuntu 18.04	Racoon, ipsec-tools, iperf, top	(8) @ 4.0 GHz	8 GB
End node B	Ubuntu 14.04	Racoon, ipsec-tools, iperf, top	(4) @ 3.2 GHz	8 GB

Table 2.2. *System configuration of SDN networks*

The core of SDN-based networks of our test bed contains 10 OpenFlow switches, five on each network (Figure 2.5). These switches are hosted on physical computers interconnected as illustrated in Figure 2.6.

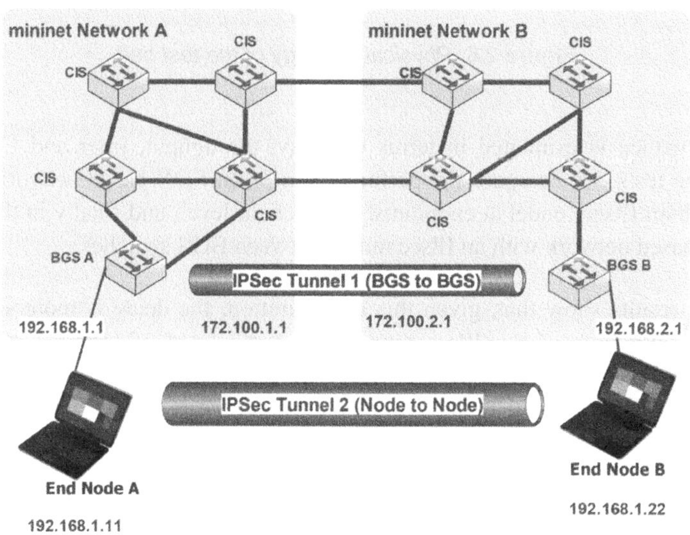

Figure 2.5. *Logical topology of the test bed*

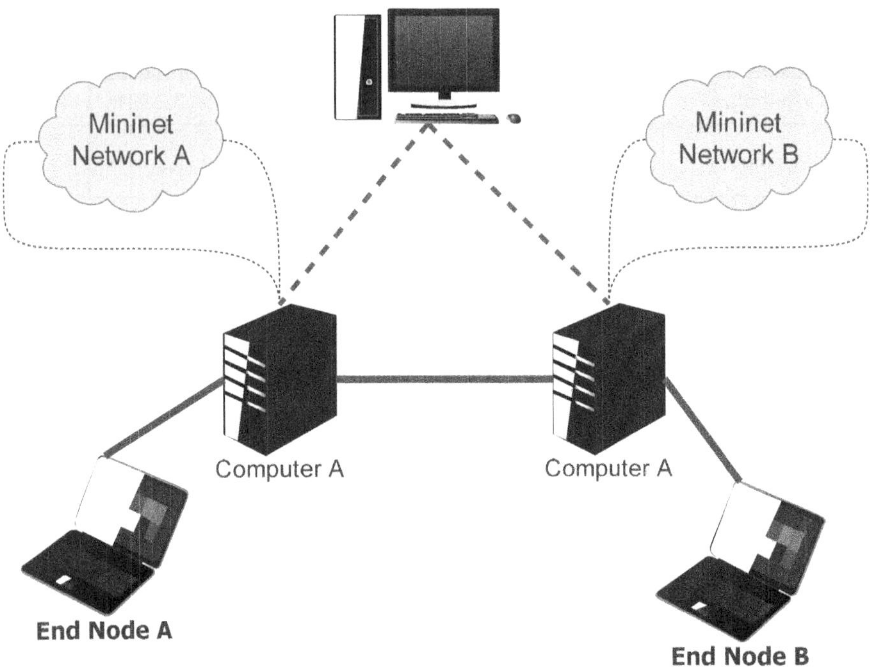

Figure 2.6. *Physical topology of the test bed*

Performance is evaluated in terms of delay, throughput, jitter and CPU load. During the tests, we compare the performances of an SDN-based network without IPsec with an IPsec tunnel at endpoints, at the client level, and finally in the case of an SDN-based network with an IPsec tunnel between BGS switches.

These results show that, given this configuration, the delay is more significant when the endpoints of the IPsec tunnel are at the level of terminal nodes than between the security gateways represented by BGS (Figure 2.7(A)(a)). We get a similar result if we look at the throughput (Figure 2.7(A)(b)), the jitter (Figure 2.7(B)(c)) and the process load (Figure 2.7(B)(d)).

Although the scope of these results should be confirmed, they can be an argument in favor of deploying security services at gateways (as opposed to terminal nodes). With this type of result, security could be managed as a service at the controller level.

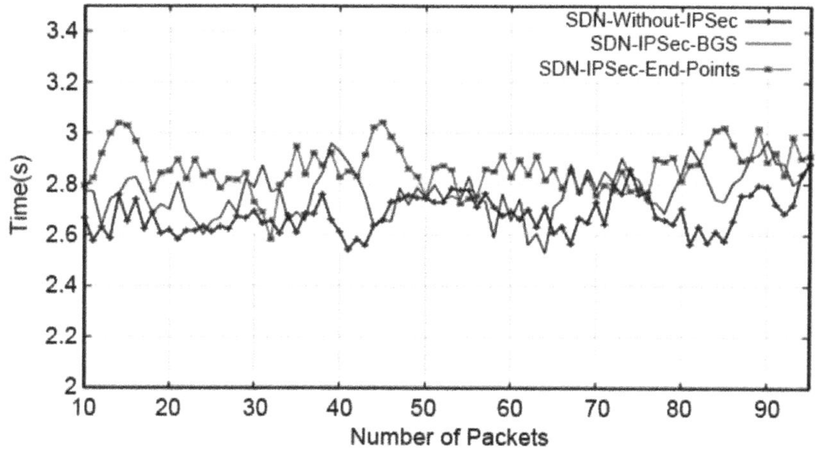

a) Performance in terms of delay

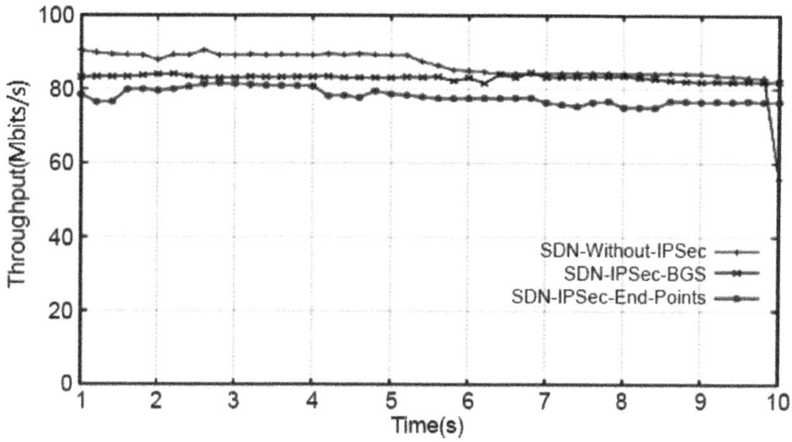

b) Performance of the SDN security service in terms of flow

Figure 2.7. (A) *Performance of the deployment of IPSec security service on SDN. For a color version of this figure, see www.iste.co.uk/benmammar/network.zip*

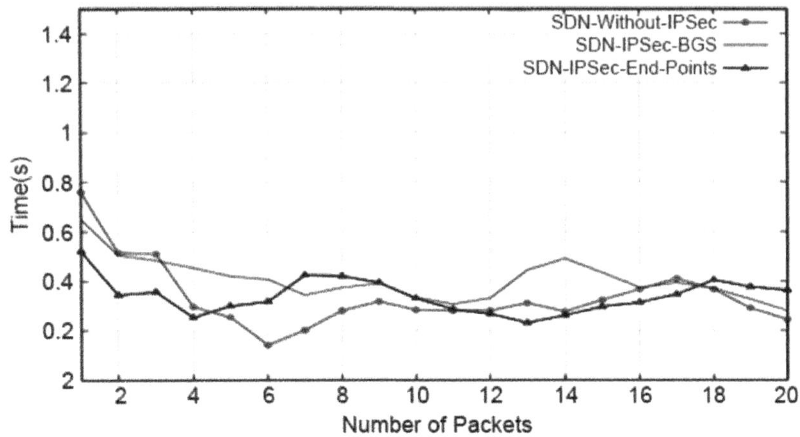

c) Performance of the SDN security service in terms of jitter

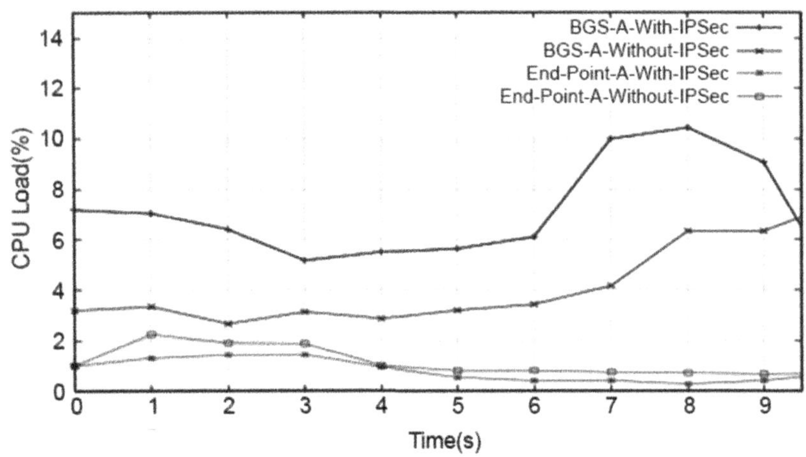

d) Performance of the SDN security service in terms of CPU load

Figure 2.7. (B) *Performance of the deployment of IPSec security service on SDN. For a color version of this figure, see www.iste.co.uk/benmammar/network.zip*

2.4. Intelligence in SDN-based networks

Intelligence-oriented design in networks is not new, but it is a challenge with significant stakes. This section starts by presenting the knowledge plane proposed in the 2000s to enable computer networks to make decisions, particularly related to management, in an "autonomous" manner (Krief 2010). We then introduce the notion of knowledge-defined networking (KDN), which must enable autonomous control and use of SDN-based networks. In the end, we present the very similar notion of intelligence-defined networking (IDN) defined by intelligence or intent-driven networks, a new paradigm brought by Huawei, which is a leader in this market.

2.4.1. *Knowledge plane*

In 2003, Clark *et al.* (2003) proposed a new plane enabling knowledge management in a network: the knowledge plane. The main objective of this new plane was to solve the Internet-specific limits, particularly the lack of reliability and adaptation when facing a new situation such as a security attack (Mbaye and Krief 2009). This work has been subsequently integrated in the architectures of autonomous networks. An autonomous network is defined as one with the capacity of management and underlying infrastructure processes to deploy, organize and operate without external aid. The administrator only guides these processes, setting high level objectives (Krief 2010). The knowledge plane plays an essential role in autonomous networks, enabling each autonomous entity to close the network control loop without administrator support. Management tasks can therefore be conducted by the network itself, each autonomous entity being capable of self-configuration, self-optimization, self-protection and self-repair. Movahedi *et al.* (2012) compared the various architectures of autonomous networks and underlined the role of learning in enabling intelligent adaptation and converging to optimal network operation.

In the context of SDN-based networks, the clear separation between the control and data planes and the centralization of intelligence can positively influence the development of the knowledge plane with new functionalities.

2.4.2. *Knowledge-defined networking*

Mestres *et al.* (2017) studied the reasons why AI techniques were not adopted in practice and stated that the expansion of the two recent paradigms, namely the SDN and the network analysis (NA), will facilitate the adoption of these techniques for network control and use. They also proposed a new paradigm, known as KDN, which associates SDN, NA and ML to provide automated network control. In Mestres *et al.* (2018), they explored the possibility of applying various ML models and techniques for modeling complex network elements, such as the virtual network function (VNF). They also proved that the behavior of various VNFs could be learnt using ML techniques, such as their CPU consumption depending on input traffic.

The ALLIANCE project (Careglio *et al.* 2018) relies on the new KDN paradigm. Its objective is to design and implement a 5G network infrastructure that is capable of dealing with ubiquitous services, while meeting the performances and commercial demands of multiple stakeholders. This project is expected to propose, compare and interconnect three different prototypes of complete network architecture (from access network to core network), autonomous and 5G oriented. Several network solutions will be studied, such as SDN/NFV, programmable overlay networks and Recursive InterNetwork Architecture (RINA). Each of the retained solutions relies on a KDN orchestrator, which takes advantage of ML techniques for automated network deployment, use, monitoring and repair.

In Hyun and Hong (2017), the authors present an autonomous network architecture combining network telemetry, KDN, SDN and P4 INT. P4 INT (In-band Network Telemetry) enables the gathering of network telemetry data. These data are then used by the KDN to bring intelligence to network management. SDN is then in charge of network management and control depending on the decision made by the knowledge plane. In Hyun *et al.* (2018), the authors present a first implementation of the network monitoring system.

In Lu *et al.* (2019), the authors take an interest in the orchestration at KDN level in order to put into operation a hybrid optical/electronic DCN in a high-performance and energy-efficient manner. This orchestration relies on three AI modules based on deep learning, namely the traffic prediction module, the module for the prediction of the virtual machine demand and the module for network reconfiguration. The experimental results show that, besides an improvement in service provision performances, this approach enables better energy efficiency in the system.

2.4.3. *Intelligence-defined networks*

Intelligence-defined network (IDN) approaches come as a development of SDN by adding a cognitive layer above the control layer. This also enables the introduction of AI in the network management and control. This approach is quite similar to KDN approach, but it is proposed by manufacturers, particularly by Huawei (Jiang 2016; Huawei 2018).

The latter proposed at the Mobile World Congress in 2018 a new offer referred to as an intent-driven network (IDN), aimed at upgrading the SDN-based networks into networks “driven by intent” by the addition of AI, big data and cloud technologies. Emphasis is on the capacity of the control plane to know – or to predict – the type of application/service that a client would want to use and the context in which he/she would want to do it, in order to automatically book for him/her the full bandwidth he/she needs, and provide him/her the best user experience. The network, now centered on the user, is thus capable of accurately identifying the client intent and fulfilling it. It is also capable of detecting in real time the quality of the user experience and conducting a predictive analysis in order to proactively optimize performances (Huang *et al.* 2018).

2.5. AI contribution to security

This section starts by presenting ML techniques that are most commonly used for security purposes. This is followed by an example of intrusion detection systems as an illustration of AI contribution to the SDN security field.

2.5.1. *ML techniques*

ML is a subdomain of AI; it can be further divided into four groups:

1) supervised learning: given a set of objects, each of which has an associated target value, the system must learn a model that is capable of predicting the proper target value of a new object;

2) unsupervised learning: given a set of objects that have no associated target value, the system must learn a model that is capable of extracting regularities in the set of objects in order to understand the data structure;

3) semi-supervised learning: given a small set of objects, each of which has an associated target value and a larger set of objects without target value, the system must be capable of solving supervised and/or unsupervised problems;

4) reinforcement learning: given a set of decision sequences in a dynamic environment and for each action of a sequence a reward value, the system must learn a model capable of predicting the best decision.

Researches in the field of computer security use mainly supervised and unsupervised learning techniques. These techniques are presented in the next sections.

2.5.1.1. *Supervised learning techniques*

Supervised learning involves a set of well-established techniques that are used in various domains.

The most commonly used in the security research field are the following:

1) *decision tree*: a decision-aiding tool representing a set of choices in the shape of a tree. The tree has nodes and branches. The nodes are assigned labels by the attributes and the branches are assigned labels by a predicate that applies to the parent node. Decision trees are very successful in solving classification problems. The decision tree can describe a set of data by a tree structure (Negnevitsky 2005). The input and output data can be discrete or continuous. The decision trees can represent all the Boolean functions. A decision tree conducts a sequence of tests, each internal node of the tree corresponding to a test of one of the input attributes (Russell and Norvig 1995). Learning involves recursive tree building by choosing the attribute that assigns the examples in their own classes to the child nodes for each value of the chosen attribute. The decision tree is a tool used in several fields, such as security and data mining;

2) *Bayesian network*: a probabilistic graphical model representing random variables as a graph. This graph represents the relations, which are not deterministic, but probabilistic, causal, between the variables. Bayesian networks involve taking simultaneously into account *a priori* knowledge (in the graph) and the experience contained in the data;

3) *genetic algorithm (GA)*: one of the best known and most commonly used techniques (Holland 1975). Here, the solution to an optimization problem is represented by a chromosome. A set of chromosomes forms the population. The two fundamental but very important operations of this technique are: crossover and mutation. The crossover operation combines previously selected individuals by exchanging some of their parts. On the other hand, mutation introduces a slight uncertainty in the search in order to avoid the local *optima* problem. The significant factors for the implementation of the whole GA are: the selection strategy and the type of crossover and mutation operators (Boussaid *et al.* 2013);

4) *support vector machine (SVM)*: these are a set of supervised learning techniques intended to solve discrimination and regression problems. SVMs were developed in the 1990s and were adopted due to their capacity to work with large size data. SVMs rely on linear separators that maximize the margin between two different classes in order to provide the best classification. The learning methods can be used to transform the input data into a large size space in order to deal with linearly nonseparable cases (Nguyen and Armitage 2008);

5) *neural networks*: artificial neural networks drew their inspiration mainly from the biological neurons of the human brain (Negnevitsky 2005). Neural networks are widely used because of their capacity to deal with large amounts of information and their stability when exposed to noise. Neural networks are a set of neurons (small processors) working in parallel. These neurons receive data, apply a function known

as an "activation function" and send them over to other neurons or to an external source. Neural networks have many advantages. First of all, they can adjust to data without explicitly specifying the function or distribution for the representation of the underlying model (Zhang 2000). Second, neural networks form a universal functional approximator, which can approximate any function (Zhang 2000). Third, neural networks are nonlinear models, which gives them the flexibility needed for the representation and modeling of complex relations (Zhang 2000). MultiLayer Perceptrons (MLP) are the most commonly used systems for neural network classification. MLPs are mainly formed with supervised learning algorithms. Neural networks are subjected to overadjustments when too many parameters are used in the model (Russell and Norvig 1995);

6) *random forest*: this technique belongs to the model aggregations family. It is a particular case of bagging (bootstrap aggregation) applied to decision trees. The principle is to average the predictions of several independent models in order to reduce variance, and therefore the prediction error. To build these various models, several bootstrap samples are selected, in other terms, unrestricted random sampling is performed. Random forests add chance at the variable level, besides the bagging principle. For each tree, a bootstrap sample of individuals must be selected and a tree node is built on a subset of randomly drawn variables;

7) *K-nearest neighbors (K-NN)*: a classification algorithm. The idea is to vote the nearest neighbors through an observation x. The class of x is determined based on the majority class among the K-NN of the observation x. The nearest neighbor is determined using a distance function.

2.5.1.2. *Unsupervised learning techniques*

The unsupervised approach also involves a set of well-established algorithms:

1) *K-means* is one of the best known clustering approaches. Its implementation requires knowing the k parameter beforehand, which indicates the number of resulting clusters. Each data point is assigned to the centroid that is closest to each group. K-means minimizes an objective function, which represents the distance between the data points and their corresponding centroids (Khan and Ahmad 2004). Centroids are updated repeatedly, depending on the data points attributed to them until they stay the same or no point changes. K-means mainly depends on the initial set of classes. Consequently, an inappropriate choice of k may lead to mediocre results (Zhang and Xia 2009). Moreover, the fuzzy-C-means (FCM) classification (Pal and Bezdek 1995), also known as "soft K-means", enables each data point to belong to several classes. In other terms, a data point may belong to all the classes with a different adhesion degree;

2) *self-organizing map (SOM)*: used in DoS attacks, the SOM algorithm is also known as the Kohonen algorithm. It is a stochastic classification algorithm that integrates a notion of vicinity between classes.

These various learning techniques are summarized in Table 2.3.

ML Algorithm	Problem	Advantages	Drawbacks
Decision tree	Classification Regression	Easy to implement Easily understandable decisions Data classification without heavy computation Continuous and discrete data processing	Costly in terms of computation and memory space Instability: a small change in the dataset may drive significant modifications in the model
Bayesian network	Classification	Easy to implement Rapid learning phase	Difficulty in managing continuous data
Genetic algorithm	Classification	Conclude proper classification rules	Computation cost
Support vector machine	Classification Regression	Big data processing	Less effective on a noised dataset Computation costly
Neural network	Classification Regression	Rapid prediction phase Works well on large volume of training data	Computation cost Difficult interpretation of the learning model by a human
Random Forest	Classification Regression	Works well on big learning data Reduces model instability compared to decision trees	Low training speed
K- NN	Classification Regression	Easy to implement	Costly in terms of computation and memory space
K-means	Clustering	Easy to implement Easy to interpret results	Cost of linear computation with the size of learning data
SOM	Clustering	Manipulation of very large volume data	Computation costly

Table 2.3. *The most commonly used machine learning techniques for computer security purposes*

2.5.2. Contribution of AI to security service: intrusion detection

The above-mentioned AI techniques have been used to solve many security problems in general, and particularly for intrusion detection (Soheily-Khah *et al.* 2018).

In Barapatre *et al.* (2008), an approach with a multilayer perceptron–back propagation (MLP-BP) is proposed. The input to the proposed system has the characteristics of the Knowledge Discovery from Database (KDD)[6] dataset and the output the classification of normal packets and suspect packets present in the dataset. The MLP-BP neural network has been shown to detect DoS and "Probe" attacks more accurately than the user to root (U2R) attacks. In Lu *et al.* (2015), the authors use the radial basis function (RBF) neural networks, which are very practical in intrusion detection systems. They compared RBF and MLP-BP using a KDD dataset that was processed by the conversion of all the chains into figures, thus reducing the size of the dataset. The simulation results showed that the RBF neural network is better than MLP-BP in terms of learning time, accuracy and detection of attacks.

Canbay and Sagiroglu (2015) propose a hybrid approach for attack detection. GA and K-NN were used in combination for attack modeling and detection. K-NN was used for the classification of the attacks and GA for the selection of *k* neighbors in a sample of attacks. This hybrid system was applied for the first time in the field of intrusion detection. The results showed that the proposed system yields better results in terms of detection accuracy than a single system.

Most researchers used the KDD dataset, which was widely criticized for not being a faithful representation of the network. In Sahu and Mehtre (2015), the authors used a new set of labeled network data, referred to as the "Kyoto 2006+ dataset". In Kyoto 2006+, each instant is labeled as "normal" (no attack), attack" (known attack) and "unknown attack". The users used the Decision Tree (J48) algorithm to classify the network packet that could be used for a Network IDS (NIDS). The results showed that the decision tree has a very good classification accuracy and that it also enables the classification of unknown attacks.

As these several work samples prove, the potential of AI has not yet been exhausted in the security field, since a combination of techniques may yield better results.

6 Available at: http://nsl.cs.unb.ca/NSL-KDD/.

2.6. AI contribution to security in SDN-based networks

This section takes intrusion detection systems as examples to illustrate the potential of AI in the field of SDN security.

Network administrators implement IDS to avoid intrusion attacks and apply the network security policy. An IDS monitors the traffic and sends intrusion alerts to the administrator console when a suspicious message is detected.

In Tang *et al.* (2016), the authors propose a deep learning approach for the detection of anomalies in an SDN environment. They built a model of deep neural network (DNN) for an intrusion detection system. They proved that deep learning has a strong use potential for anomaly detection in SDN environments. In Tang *et al.* (2018), the authors improve the approach presented in Tang *et al.* (2016). They propose an intrusion detection system that uses Gated Recurrent Unit Recurrent Neural Network (GRUN-RNN) for SDN networks. The proposed approach is tested using a NSL-KDD[7] dataset. Experimental results show that the proposed approach has also a strong potential for intrusion detection in SDN environments. Thaseen and Kumar (2013) aimed to evaluate various tree-based classification algorithms that classify the network events in intrusion detection systems. Experiments are conducted on the NSL-KDD dataset. The results show that the RandomTree model has a better accuracy if false positives are minimized.

In Chen and Yu (2016), the authors propose a new collaborative intrusion prevention architecture (CIPA). A CIPA is deployed as virtual network of a neural network on the subnetwork. Taking advantage of the parallel and simple mathematical handling of neurons in a neural network, CIPA can spread its light computation capability to the programmable switches in the subnetwork. Each programmable switch virtualizes one to several neurons. The whole neural network operates as an integrated IDS/IPS. This enables CIPA to detect the attacks distributed in a global view, which does not require extended communication and computation time.

The DDoS attack is one of the most recurrent problems in network security. Despite the fact that DDoS attack mechanisms are widely understood, the problems have recently become increasingly frequent due to the similarity between a DDoS attack and normal traffic. In Nam *et al.* (2018), the authors propose approaches for the detection of DDoS attacks. The algorithms proposed in the detection architecture are implemented in an SDN environment. The SDN controller enables the rapid compilation of a complex classification and detection algorithm. Experimental

7 Available at: http://nsl.cs.unb.ca/NSL-KDD/.

results show that these algorithms have a relatively short processing time, while maintaining good accuracy.

In Mihai-Gabriel and Victor-Valeriu (2014), the authors present a way to reduce DDoS attacks in an SDN environment by evaluating the risks using a cyberdefense system based on neural networks.

In Niyaz *et al.* (2016), the authors propose a multivector DDoS detection system based on deep learning in an SDN environment. They implemented the system as network application above an SDN controller. Deep learning was used to reduce a large number of functionalities derived from the network traffic headers. Using the traffic traces gathered from various scenarios, the authors detected attacks with high accuracy and very few false positives.

Braga *et al.* (2010) present a light method for DDoS attack detection based on traffic flow characteristics, in which such information is extracted in a relatively short time compared to traditional approaches. This is made possible by the use of the NOX platform, which provides a programming interface facilitating the processing of switch information. The approach has good detection performance and a very low rate of false positives.

In Mehdi *et al.* (2011), the authors develop algorithms for the detection of significant traffic anomalies that can be implemented in an SDN context using Openflow and NOX compatible switches as controller. They prove that these algorithms are accurate in the identification of malicious activities in domestic networks with respect to Internet service provider (ISP). Moreover, the analysis of efficiency of SDN implementations on a programmable domestic network router indicates that anomaly detectors can operate without affecting the performances of domestic network traffic.

2.7. Deployment of an intrusion prevention service

In the context of signature-based IDS, when a network attack occurs, one of the most difficult tasks for the network administrator or expert is to write an accurate signature describing this new attack.

This section presents the works of Hamdi *et al.* (2015), who use another AI technique: inductive logical programming (Muggleton 1991) for automated signature generation.

Finally, we present a proposal for advancing this work, by enabling the "intelligent" deployment of such a security service for SDN-based networks.

2.7.1. *Attack signature learning as cloud service*

The proposal in Hamdi *et al.* (2015) is to provide attack signature learning as a Cloud service. The service is deployed in the cloud with a client/server architecture. Clients have access to the service via secured VPN.

The learning service is provided by learning nodes (LN) containing inductive logic programming (ILP) modules. The LNs receive as input malicious and normal traffics in standard formats (for example, PCAP). The system replies with a Prolog rule (signature) that can be translated into any specific target grammar.

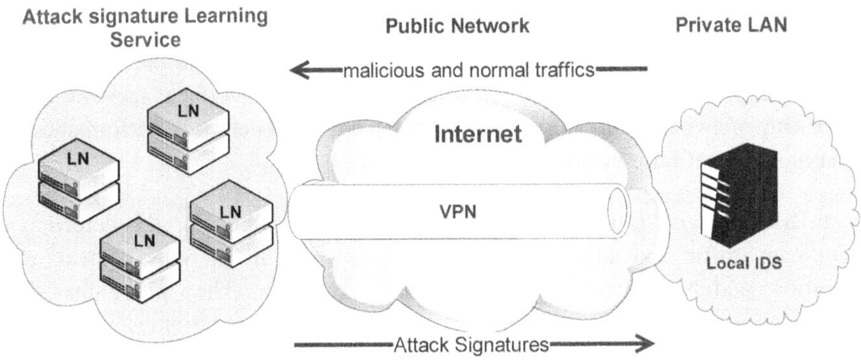

Figure 2.8. *Architecture of signature learning service*

Insiden an LN there is an ILP engine, a facts base and a grammar translator proxy (GTP). This proxy is a component of data mediation between the ILP engine and the client. It translates the input data from the client into predicates corresponding to ILP engine parameters:

– malicious traffic → positive examples (E^+);

– normal traffic → negative examples (E^-).

The ILP engine learns the rules corresponding to the attack signatures presented to it. This engine needs a facts base composed of background knowledge (BK) and examples that correspond to traffics classified as malicious.

Once the ILP engine learns the rules, GTP transforms these rules of Prolog grammar into the target IDS-specific grammar.

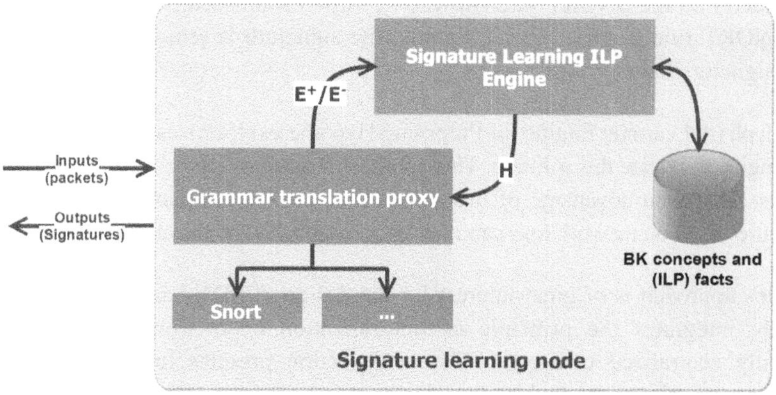

Figure 2.9. *Architecture of a Learning Node (LN)*

For the time being, only the grammar for IDS SNORT rules is taken charge of by the proposal. Let the learnt signature be $C \vDash P_1 \dots P_n$, where C is the concept at the clause heading and $P_1 \dots P_n$ are predicates of BK. The transformation of the SNORT signature rule involves two stages.

Stage 1: build the rule heading based on the Prolog signature by extracting the source and destination IP addresses, the port numbers in the learnt signature. The BK predicates that can be in the signature rule body are: `Pi=ip_src(packet)`, `Pj=ip_dst(packet)`, `Pk=src_port(packet)`, `dst_port(packet)`, `Pl=proto(X,tcp)`. Table 2.4 gives an idea of the transformation done.

(Prolog) Signature	SNORT signature
ip_src(192.168.0.1) ; ip_dst(10.0.0.1),proto(P,tcp) ; dst_port(X),src_port(Y)	alert tcp 192.168.0.1:ANY > 10.0.0.1:ANY
ip_src(X),dst_ip(10.0.0.1), proto(P,icmp)	alert icmp ANY > 10.0.0.1

Table 2.4. *Construction of rule heading*

Stage 2: build the body of rules. In the SNORT body, C is transformed into (msg: some_text_id). The body predicates $P_1 \ldots P_n$ are transformed in the closest SNORT filter F_1, F_2, F_3, etc. For example, `syn_tcp_actif` (packet) could be translated into the SNORT filter flow: (…, Flow:Established,…). The final body of the SNORT rule is: (F1, F2,…). Finally, the signature is sent to the client to feed their signature base.

Aleph (A Learning Engine for Proposing Hypotheses) (Srinivasan 2000) was used as ILP engine to realize this solution. This approach was shown to be very effective in DoS attacks. The main advantage of this solution is that the complex task of writing the signature based on network logs can now be accomplished by the cloud service.

This approach is of great interest for the design of SDN-based networks, which already integrates the principle of delocalization of network intelligence in a logically centralized controller. The next section presents the proposals on the architectural plane to enable the deployment of such services in SDN-based networks.

2.7.2. *Deployment of an intrusion prevention service in SDN-based networks*

One of the first challenges encountered when deploying an IDS architecture that learns the signatures on-the-fly and deploys them is the acceptable response time. The language filter systems of IDS are traditionally much richer than Southbound API in SDN networks. It is therefore a challenge to implement an intrusion prevention system (IPS) with an acceptable response time.

The objective of the proposed architecture is to provide a self-managed IPS for SDN networks. The IPS enables the extension of the security functions from one SDN controller through a P4 interface (Bosshart *et al.* 2014) because of AI. In fact, the controller's task is to transfer the retransmission rules or ACL (Access Lists) to switches supporting OpenFLow inside the network. Nevertheless, it is not designed to analyze and detect the attacks inside the network. IDS and IPS are the tools dedicated to this task. The implementation of these tools is a real challenge, given the large amount of data circulating in the SDN domain and the performance constraints related to virtualization. This architecture approaches this challenge by means of three operating blocks: a signature-based intrusion detection and prevention system, a system for on-the-fly learning of attack signatures based on ILP and deep learning, and a P4 interface enabling the deployment of new detection rules in the network.

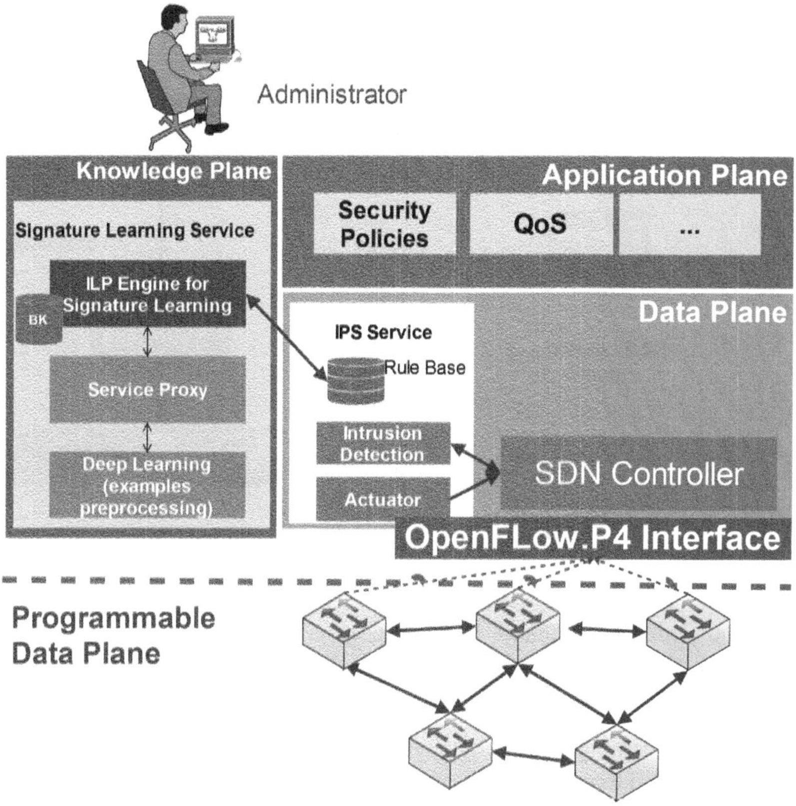

Figure 2.10. *Architecture of an intelligent SDN with IDS*

The system (Figure 2.10) operates as a processing chain that is self-optimizing with the occasional intervention of the administrator as oracle when needed. The data plane has a classical role in this architecture; on the other hand, the control plane demonstrates more intelligent activity through the knowledge plane.

In SDN-based networks, the knowledge plane manages the whole knowledge in order to improve/optimize the controller operation to ensure the security functions (policy optimization, intrusion prevention, detection of malicious operation of a switch, etc.), routing (route optimization, self-healing of disappeared routes, prediction of load distribution, etc.) and QoS.

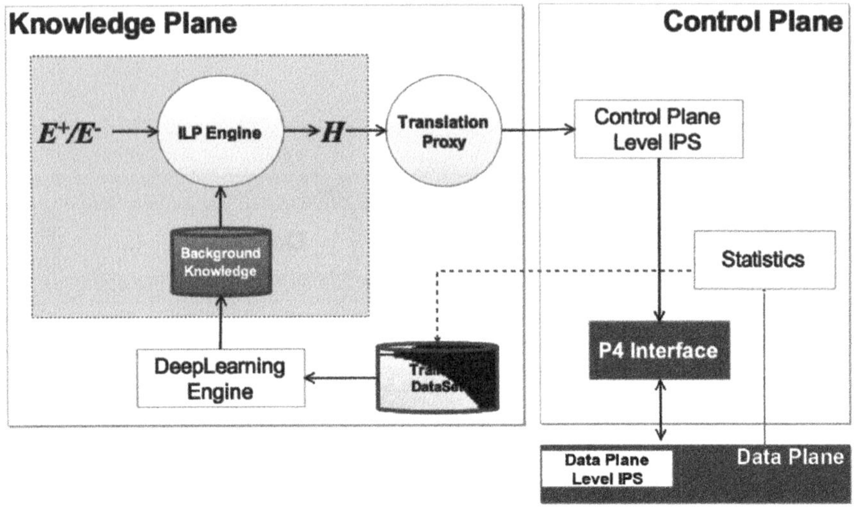

Figure 2.11. *Interactions between components of data architecture*

The knowledge plane makes it possible to have a local intelligent loop for the management of a signature-based IPS service. The loop is described in Figure 2.11. In this loop, there are two learning tools with connected roles:

– the main learning tool relies on ILP. ILP makes it possible to learn a hypothesis from examples of malicious packets as indicated in Hamdi *et al.* (2015);

– the deep learning module that, based on examples of a reference dataset such as NSL-KDD, builds a BK that makes it possible to consider only those concepts that are relevant for learning an attack:

$$\Delta(D_{set}, E) = \{p \in P \,/\, \Pi(p, D_{set}) > 0\} \quad [2.1]$$

where Δ is the deep learning function that takes as parameter a dataset (such as NLS_KDD) and positive examples of the attack. The output of this function must be the set of p properties that are relevant for BK. Relevance is represented by the function $\Pi(p, D_{set}) > 0$. This formula expresses the fact that the system will use a training base to determine the fields of parameters that are potentially relevant to appear in the attack signature. Indeed, the BK of ILP module generally contains predicates that can be used to describe the attack. The choice of BK content is crucial for having the most accurate possible rules. To add bias to BK design, we use deep learning in order to determine for each attack the properties that are most susceptible to appear in BK. The deep learning module uses the bases of training traces, in our case NSL-KDD, to optimize the BK content. This optimization involves mainly the reduction of the size that enables the upstream filtering of the most relevant parameters for an attack description, and therefore the reduction of the research space of the ILP engine.

A P4 interface makes it possible to transform the filters and actions of the intrusion detection system into programs that can be deployed in programmable switches.

The intrusion detection service presented in this section is deployed by means of a single SDN controller. To allow for scalability, we recommend the existence of physically distributed but logically centralized controllers, as well as a flat model, a solution that seems to be widely accepted at present. For exchanges between controllers in order to maintain the global view, and in the absence of standardized Eastbound/Westbound API, the Pub/Sub approach should also be retained, as it makes it possible to keep a controller informed only during a change of state.

2.8. Stakes

The stakes of using AI and ML tools for security management in SDN-based networks are very high. This is true not only in technical terms, but also from a societal perspective. Indeed, computer security has today an impact on everyday life, especially with the coming 5G. This technology will significantly impact several domains such as medicine, health, road safety, etc.

First of all, with the advent of big data and big analytics, the devices that enable the study of system vulnerabilities are increasingly effective (Terzi *et al.* 2017; Hajizadeh *et al.* 2018; Hoon *et al.* 2018). If such devices start to identify security vulnerabilities that are intrinsic to the design of SDN, and no protection is possible, the potential consequence is to simply give up this technology. It is therefore important that SDN controllers feature an intelligence and autonomy enabling them to get reprogrammed and reprogram the data plane in an adapted manner.

On the other hand, intelligent attack systems may lead to the appearance of vulnerabilities at the limit of normality in intrusion detection systems based on traditional anomalies, which can generate many false positives and negatives. But given a large number of false positives, the administrators may be tempted to slightly relax the security or to integrate exceptions for actual attacks. Indeed, when faced with intelligent attacks, the most adapted defense is another AI that is able to support the administrators in making the proper decisions.

The coming of 5G highlights the use of AI in security tools even more. Indeed, the large amount of data cannot be analyzed with classical technologies and tools. On the one hand, the security tools may be unable to provide a response within acceptable time: an attack might be successful while the detection system provides a response. On the other hand, these tools may create bottlenecks if the system operates in paranoid mode examining all the packets. The AI/ML tools may enable the limitation of analysis to traffics that are relevant for analysis. If this amount of data could have a linear behavior with respect to the traffic increase, security services could be deployed while ensuring scalability.

2.9. Conclusion

The concept of an SDN-based network was created in order to meet the challenges related to network development. The basic principles of SDN-based networks are as follows: introduction of transmission equipment programmability via standard and open Southbound interfaces; decoupling the control plane and the data plane and the global view of the network by a logical centralization of "network intelligence".

With this new approach, the elements of the data plane are programmable and very effective in packet transmission, as they conduct less complex tasks. The components of the control plane are implemented on a controller that concentrates the network driving activities due to its global view.

Nevertheless, the security question is an important stake in the development of SDN-based networks. Indeed, the weak points represented by the controller and the securing of programmable network equipment in an environment generating large amount of data are the main challenges to be met.

2.10. References

Abubakar, A. and Pranggono, B. (2017). Machine learning based intrusion detection system for software defined networks. *Seventh IEEE International Conference on Emerging Security Technologies (EST)*. 6–8 September 2017, Canterbury, UK, 138–143.

Bannour, F., Souihi, S., and Mellouk, A. (2018). Distributed SDN control: Survey, taxonomy, and challenges. *IEEE Communications Surveys & Tutorials*, 20(1), 333–354.

Barapatre, P., Tarapore, N.Z., Pukale, S.G., and Dhore, M.L. (2008). Training MLP neural network to reduce false alerts in IDS. *International Conference on Communication and Networking*, 1–7.

Bi, J., Zhang, K., and Cheng, X. (2009). Intrusion detection based on RBF neural network. *International Symposium on Information Engineering and Electronic Commerce*, 357–360.

Boussaid, I., Lepagnot, J., and Siarry, P. (2013). A survey on optimization meta-heuristics. *Information Sciences*, 237, 82–117.

Braga, R., Mota, E., and Passito, A. (2010). Lightweight DDoS flooding attack detection using NOX/OpenFlow. *35th Conference on Local Computer Networks (LCN)*, 408–415.

Canbay, Y. and Sagiroglu, S. (2015). A hybrid method for intrusion detection. *14th International Conference on Machine Learning and Applications (ICMLA)*, 156–161.

Canini, M., De Cicco, D., Kuznetsov, P., Levin, D., Schmid, S., and Vissicchio, S. (2014). STN: A robust and distributed SDN control plane. *Proceedings of Open Networking Summit (ONS'14)*. March 2–4 2014, Santa Clara, CA, USA, 1–2.

Careglio, D., Spadaro, S., Cabellos, A., Lazaro, J.A., Perelló, J., Barlet, P., Gené, J.M., and Paillissé, J. (2018). ALLIANCE Project: Architecting a knowledge-defined 5G-enabled network infrastructure. *20th International Conference on Transparent Optical Networks (ICTON)*.

Chen, X.F. and Yu, S.Z. (2016). CIPA: A collaborative intrusion prevention architecture for programmable network and SDN. *Computers & Security*, 58, 1–19.

Chung, S.P. and Mok, A.K. (2006). Allergy attack against automatic signature generation. *9th International Conference on Recent Advances in Intrusion Detection*.

Clark, D.D., Partridge, C., Ramming, J.C., and Wroclawski, J.T. (2003). A Knowledge plane for Internet. *Conference on Applications, Technologies, Architectures, and Protocols for Computer Communications*, 3–10.

Coly, A. and Mbaye, M. (2019). S-SDS: A framework for security deployment as service in software defined networks. *Third EAI International Conference*.

Cordeiro, W.L.D.C., Marques, J.A., and Gaspary, L.P. (2017). Data plane programmability beyond openflow: Opportunities and challenges for network and service operations and management. *Journal of Network and Systems Management*, 25(4), 784–818.

Das, S. and Nene, M.J. (2017). A survey on types of machine learning techniques in intrusion prevention systems. *International Conference on Wireless Communications, Signal Processing and Networking (WiSPNET)*, 2296–2299.

Dey, S.K., Rahman, M.M., and Uddin, M.R. (2018). Detection of flow based anomaly in OpenFlow controller: Machine learning approach in software defined networking. *4th International Conference on Electrical Engineering and Information & Communication Technology*.

Erickson, D. (2013). The beacon openflow controller. *Proceedings of the Second ACM SIGCOMM Workshop on Hot Topics in Software Defined Networking*.

Fortes, J. (2013). Cloud computing security: What changes with software-defined networking? *ARO Workshop on Cloud Security*.

Gowtham, V.N., Baratheraja, R.N., Jayabarathi, G., and Vetriselvi, V. (2018). Collaborative intrusion detection system in SDN using game theory. In *Proceedings of the International Conference on Computing and Communication Systems*, Mandal, J., Saha, G., Kandar, D., and Maji, A. (eds). Springer, Singapore.

Hajizadeh, M., Phan, T.V., and Bauschert, T. (2018). Probability analysis of successful cyber attacks in SDN-based networks. *Conference on Network Function Virtualization and Software Defined Networks (NFV-SDN)*, 1–6.

Hamdi, O., Mbaye, M., and Krief, F. (2015). A cloud-based architecture for network attack signature learning. *Conference on New Technologies Network and Security*.

Holland, J.H. (1975). *Adaption in Natural and Artificial Systems*. University of Michigan Press, Ann Arbor.

Hoon, K.S., Yeo, K.C., Azam, S., Shunmugam, B., and De Boer, F. (2018). Critical review of machine learning approaches to apply big data analytics in DDoS forensics. *International Conference on Computer Communication and Informatics*, 1–5.

Huang, C.H., Lee, T.H., Chang, L.H., Lin, J.R., and Horng, G. (2018). Adversarial attacks on SDN-based deep learning IDS system. *International Conference on Mobile and Wireless Technology*, 181–191.

Huawei (2018). Roads to a better future. *Mobile World Congress*.

Hyun, J. and Hong, J.W.K. (2017). Knowledge-defined networking using in-band network telemetry. *19th Asia-Pacific Network Operations and Management Symposium*.

Hyun, J., Van Tu, N., and Hong, J.W.K. (2018). Towards knowledge-defined networking using in-band network telemetry. *IEEE/IFIP Network Operations and Management Symposium*.

Jiang, S. (2016). Intelligence Defined Network IDN. *IETF 97*.

Karakus, M. and Durresi, A. (2017). A survey: Control plane scalability issues and approaches in software-defined networking (SDN). *Computer Networks*, 112, 279–293.

Kaur, S., Singh, J., and Singh Ghumman, N. (2014). Network programmability using POX controller. *International Conference on Communication, Computing and Systems*.

Kent, S. and Seo, K. (2005). Security Architecture for the Internet Protocol. Memo, Network Working Group, Obsoletes 2401.

Khan, S.S. and Ahmad, A. (2004). Cluster center initialization algorithm for K-means clustering. *Pattern Recognition Letters*, 25(11), 1293–1302.

Klaedtke, F., Karame, G.O., Bifulco, R., and Cui, H. (2014). Access control for SDN controllers. *Proceedings of the 3rd Workshop on Hot Topics in Software Defined Networking*, 219–220.

Kreutz, D., Ramos, F.M.V., Veríssimo, P.E., Rothenberg, C.E., Azodolmolky, S., and Uhlig, S. (2015). Software-defined networking: A comprehensive survey. *Proceedings of the IEEE*, 103(1), 14–76.

Krief, F. (ed.) (2010). *Communicating Embedded Networks: Network Applications*. ISTE Ltd, London and Wiley, New York.

Lu, J., Hongping, H., and Yanping, B. (2015). Generalized radial basis function neural network based on an improved dynamic particle swarm optimization and AdaBoost algorithm. *Neurocomputing*, 152, 305–315.

Lu, W., Liang, L., Kong, B., Li, B., and Zhu, Z. (2020). AI-assisted knowledge-defined network orchestration for energy-efficient datacenter networks. *IEEE Communications Magazine*. IEEE, 58(1), 86–92.

Ma, L., Zhang, Z., Ko, B., Srivatsa, M., and Leung, K.K. (2018). Resource management in distributed SDN using reinforcement learning. *SPIE Defense + Security*, April 15–19.

MacQueen, J.B. (1967). Some methods for classification and analysis of multivariate observations. *Proceedings of the 5th Berkeley Symposium on Mathematical Statistics and Probability*, 281–297.

Mbaye, M. and Krief, F. (2009). A collaborative knowledge plane for autonomic networks. In *Autonomic Communication*, Vasilakos, A.V., Parashar, M., Karnouskos, S., and Pedrycz, W. (eds). Springer, Boston, 60–90.

Mehdi, S.A., Khalid, J., and Khayam, S.A. (2011). Revisiting traffic anomaly detection using software defined networking. *International Workshop on Recent Advances in Intrusion Detection*, 161–180.

Mestres, A., Alarcón, E., and Cabellos, A. (2018). A machine learning-based approach for virtual network function modeling. *IEEE Wireless Communications and Networking Conference Workshops (WCNCW)*.

Mestres, A., Rodriguez-Natal, A., Carner, J. *et al.* (2017). Knowledge-defined networking. *ACM SIGCOMM Computer Communication Review*, 47(3), 2–10.

Mihai-Gabriel, I. and Victor-Valeriu, P. (2014). Achieving DDoS resiliency in a software defined network by intelligent risk assessment based on neural networks and danger theory. *15th International Symposium on Computational Intelligence and Informatics (CINTI)*, 319–324.

Movahedi, Z., Ayari, M., Langar, R., and Pujolle, G. (2012). Survey of autonomic network architectures and evaluation criteria. *IEEE Communications Surveys & Tutorials*. IEEE publisher, 14(2), 464–490.

Muggleton, S. (1991). Inductive logic programming. *New Generation Computing*, 8, 295–318.

Nam, T.M., Phong, P.H., Khoa, T.D. *et al.* (2018). Self-organizing map-based approaches in DDoS flooding detection using SDN. *International Conference on Information Networking (ICOIN)*, 249–254.

Negnevitsky, M. (2005). *Artificial: A Guide to Intelligent Systems*. Addison-Wesley, Boston, MA.

Newsome, J., Karp, B., and Song, D. (2006). Paragraph: Thwarting signature learning by training maliciously. *Proceedings of the 9th International Conference on Recent Advances in Intrusion Detection*, 81–105.

Nguyen, T.T. and Armitage, G. (2008). A survey of techniques for internet traffic classification using machine learning. *IEEE Communications Surveys & Tutorials*. IEEE publisher, 10(4), 56–76.

Niyaz, Q., Sun, W., and Javaid, A.Y. (2016). A deep learning-based DDoS detection system in software-defined networking (SDN) [Online]. Available at: arXiv:1611.07400.

Nkosi, M., Lysko, A., Ravhuanzwo, L., Nandeni, L., and Engelberencht, A. (2016). Classification of SDN distributed controller approaches: A brief overview. *International Conference on Advances in Computing and Communication Engineering (ICACCE)*.

Oktian, Y.E., Lee, S., Lee, H., and Lam, J. (2017). Distributed SDN controller system: A survey on design choice. *Computer Networks*, 121, 100–111.

ONF (2015). Software-defined networking: The new norm for networks [Online]. White paper. Available at: https://www.opennetworking.org/images/stories/downloads/sdn-resources/white-papers/wp-sdn-newnorm.pdf, [Accessed February 2019].

Pal, N.R. and Bezdek, J.C. (1995). On cluster validity for the fuzzy c-means model. *IEEE Transactions on Fuzzy Systems*, 3(3), 370–379.

Pat, B., Dan, D., Glen, G., Martin, I., Nick McKeown, Jennifer, R., Cole, S., Dan, T., Amin, V., George, V., and David, W. (2014). P4: Programming protocol-independent packet processors. *ACM SIGCOMM Computer Communication Review*. ACM publisher, 44(3), 87–95.

Rebecchi, F., Boite, J., Nardin, P.-A., Bouet, M., and Conan, V. (2017). Traffic monitoring and DDoS detection using stateful SDN. *Conference on Network Softwarization (NetSoft)*, 1–2.

Russell, S. and Norvig, P. (1995). *Artificial Intelligence (A Modern Approach)*. Prentice Hall, Upper Saddle River.

Sahu, S. and Mehtre, B.M. (2015). Network intrusion detection system using J48 decision tree. *International Conference on Advances in Computing, Communications and Informatics (ICACCI)*, 2023–2026.

Shu, Z., Wan, J., Li, D., Lin, J., Vasilakos, A.V., and Imran, M. (2016). Security in software-defined networking: Threats and countermeasures. *Mobile Networks and Applications*, 21(5), 764–776.

Soheily-Khah, S., Marteau, P.F., and Béchet, N. (2018). Intrusion detection in network systems through hybrid supervised and unsupervised machine learning process: A case study on the ISCX dataset. *1st International Conference on Data Intelligence and Security (ICDIS)*, 219–226.

Srinivasan, A. (2000). The aleph manual. Technical Report, Oxford University, Oxford.

Tang, T.A., Mhamdi, L., McLernon, D., Zaidi, S.A.R., and Ghogho, M. (2016). Deep learning approach for network intrusion detection in software defined networking. *International Conference on Wireless Networks and Mobile Communications (WINCOM)*, 258–263.

Tang, T.A., Mhamdi, L., McLernon, D., Zaidi, S.A.R., and Ghogho, M. (2018). Deep recurrent neural network for intrusion detection in SDN-based networks. *4th Conference on Network Softwarization and Workshops (NetSoft)*, 202–206.

Terzi, D.S., Terzi, R., and Sagiroglu, S. (2017). Big data analytics for network anomaly detection from netflow data. *International Conference on Computer Science and Engineering (UBMK)*, 592–597.

Thaseen, S. and Kumar, C.A. (2013). An analysis of supervised tree-based classifiers for intrusion detection system. *International Conference on Pattern recognition, informatics and mobile engineering (PRIME)*, 294–299.

Wang, J., Wang, Y., Hu, H., Sun, Q., Shi, H., and Zeng, L. (2013). Towards a security-enhanced firewall application for openflow networks. In *Cyberspace Safety and Security. Lecture Notes in Computer Science*, Wang, G., Ray, I., Feng, D., and Rajarajan, M. (eds). Springer, Basel.

Xie, J., Yu, F.R., Huang, T., Xie, R. *et al.* (2019). A survey of machine learning techniques applied to software defined networking (SDN): Research issues and challenges. *IEEE Communications Surveys and Tutorials*, 21(1), 393–430.

Zhang, C. and Xia, S. (2009). K-means clustering algorithm with improved initial center. *Second International Workshop on Knowledge Discovery and Data Mining*, 790–792.

Zhang, G.P. (2000). Neural networks for classification: A survey. *IEEE Transactions on Systems, Man, and Cybernetics, Part C (Applications and Reviews)*, 30(4), 451–462.

Zhang, Y., Cui, L., Wang, W., and Zhang, Y. (2017). A survey on software defined networking with multiple controllers. *Journal of Network and Computer Applications*, 103, 101–118.

PART 2

AI and Network Optimization

3

Network Optimization using Artificial Intelligence Techniques

Asma AMRAOUI and Badr BENMAMMAR
Abou Bekr Belkaid University, Tlemcen, Algeria

3.1. Introduction

The telecommunications field has greatly progressed in recent years due to the booming markets of mobile phones and Internet and the deployment of broadband networks and intelligent networks. Thanks to these developments, network environments have become more complex, as they permanently process a huge amount of information; this renders network management very difficult.

Communication services providers must currently meet increasing customer demands for better quality services and better customer experience. Telecommunication companies seize these opportunities and are taking advantage of the large amount of data collected throughout the years from their broad customer base. These data are extracted from equipment, networks, mobile applications, geo-localizations and detailed customer profiles.

In order to process and analyze such huge volumes of data and extract useful information, telecommunication companies take advantage of artificial intelligence (AI) to provide better customer experience, improve operations and increase company revenues with new products and services.

Intelligent Network Management and Control,
coordinated by Badr BENMAMMAR.

Indeed, AI can help identify the anomalies and proactively solve the problems before they affect customers, and hence optimize the network. Network optimization, predictive maintenance and virtual assistants are examples of cases in which AI had an impact on the telecommunications sector.

This chapter deals with AI in general, and defines the various intelligent techniques commonly used in the telecommunications sector, including expert systems (ESs), machine learning, multiagent systems (MASs), but also the Internet of Things (IoT) and big data, which are very trendy and successful in telecommunications companies.

This chapter focuses on four aspects of network optimization: network performances, quality of service (QoS), security and energy consumption. For each of these criteria, an explanation is provided on what their optimization involves and how AI can contribute to better use.

3.2. Artificial intelligence

3.2.1. *Definition*

Human intelligence is opposed to instinct, which is associated with reflex rather than with elaborated thought. As a science, AI was founded after the Second World War, once the first electronic computers were invented. This science had a twofold objective: to simulate human capacities in an attempt to better understand human intelligence and to replace human labor in some automatic and repetitive tasks. A person's intelligence is often associated with his/her reasoning and thinking capacity.

The term AI was coined by John McCarthy and Marvin Lee Minsky, who defined it as:

> "the construction of computer programs that engage in tasks that are currently more satisfactorily performed by human beings because they require high-level mental processes such as: perceptual learning, memory organization and critical reasoning".

AI could be considered to help in the design of systems that are able to replicate human behavior in their reasoning activities.

For several years now, intelligence has almost always been associated with learning capacities. Because of learning, an intelligent system can execute a task and improve its performances with experience.

3.2.2. *AI techniques*

3.2.2.1. *Expert systems*

An expert system (ES) is a tool that can replicate the cognitive mechanisms of a human expert in a particular domain. It is one of the paths potentially leading to AI. More precisely, an ES is a software that can answer questions by means of reasoning based on facts and known rules.

ESs are generally composed of:

– knowledge base;

– interface;

– inference engine.

The knowledge base is a set of data that are used by the inference engine. It stores the field-specific knowledge of the system. It gathers all the knowledge of an expert in the respective field.

The knowledge base contains:

– *engagement standards (expert knowledge)*: basic information and system configuration information, measures, laws, parameters and contractual data;

– *inference rules (know-how)*: set of logic deduction rules used by the inference engine;

– *facts (experience) base*: set of data based on which the system starts to operate. This base is enriched as the system makes deductions. This work space is a sort of short-term memory, where the system also stores pending rules, subproblems, etc.

Interfaces are used for the dialogue between the expert, in charge with creating the knowledge base, and the machine.

The inference engine is the mechanism that enables the inference of new knowledge from the system's knowledge base. It is the system's brain and it is used for triggering the rules and chaining them one after the other.

The two most commonly employed mechanisms for triggering rules are:

– forward chaining;

– backward chaining.

An ES is different from classical software. Indeed, classical software is developed around a set of algorithmic processes. Problem resolution follows a sequence of stages that is well defined by the programmer. An ES can integrate the capacity to determine by itself the processes adapted to a given state of the input parameters, in other terms a sequence of stages that was not predefined by the programmer for this state. This difference between classical software and an ES is essentially due to the method of organization and use of specialized knowledge.

The main advantage of an ES is given by its very high performances in solving the problems encountered during the expertise period for which rules were formulated. Nevertheless, for a large size domain, the number of rules increases significantly and their maintenance is increasingly difficult. Indeed, it should be possible to carry on the expertise of the studied domain, formulate new rules and manually correlate them with all existing rules.

An ES is therefore highly adapted to fields that change to a very little extent. On the other hand, if the field is very dynamic, certain expert rules may very rapidly become obsolete and weaken the system, rendering it unable to solve certain problems. These weaknesses of rule-based ESs led to the development of a new approach to the representation of expert knowledge.

There are two types of ESs:

– rule-based classical ESs, which formulate rules in order to describe and understand the propagation of faults and alarms in a telecommunication network;

– model-based evolutionary ESs, which draw their inspiration from AI sciences and consider a phenomenon as understood only if it can be replicated or simulated. This category includes model-based diagnostic methods that develop reasoning on an explicit representation of the network structure and operation and the methods that try to artificially learn the network behavior without modeling it.

3.2.2.2. *Case-based reasoning*

Case-based reasoning (CBR) is a paradigm of AI, which involves solving a new problem, referred to as "target problem" based on a set of already solved problems. CBR is analogical reasoning that globally satisfies what is known as the "analogy square", as illustrated in Figure 3.1.

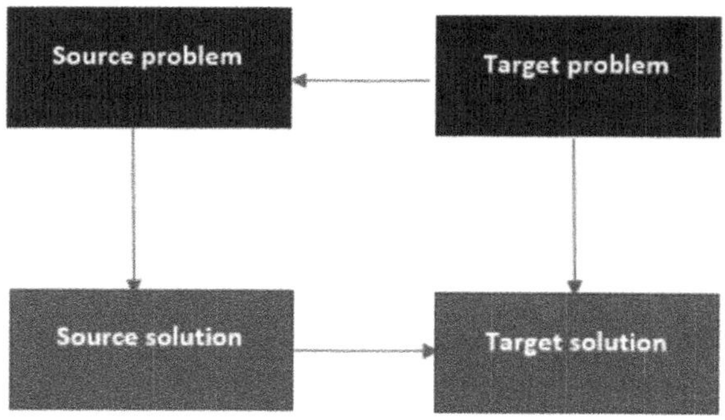

Figure 3.1. *Analogy square of ES*

The research of similar source cases is naturally essential in the cycle. Let us recall that the source case to be chosen is normally the case whose problem description is the closest possible to the description of the target problem.

Reuse involves reusing a similar case, in order to have a reasoning trace of the target case, while the review enables a correction, so that the case has a correct solution.

Learning a new solved case is an opportunity for the knowledge base to become enriched.

The principle of CBR involves the recovery, adaptation and execution of the solutions to previous problems in order to evaluate current problems. The past diagnostic solutions are stored as cases in a knowledge base. The cases contain the most relevant characteristics of past diagnostic solutions; they are adapted and used to solve the new problems.

The experience acquired through the diagnosis of these new problems constitutes new cases stored for future use. This system integrates the capacity to learn not only from its previous correct diagnostic solutions, but also from its failures. Indeed, when the attempt to diagnose a situation fails, the system identifies and logs the reason of this failure, so that it can remember it during future diagnoses.

A CBR system has a case base. For each case, there is a detailed description of the problem and a solution. Moreover, an engine is needed in order to use this information. The engine finds the cases that are similar to the new problem to be solved. After analysis, the engine provides an adapted solution that must be validated. Finally, the engine adds the problem and its solution to the case base.

3.2.2.3. *Machine learning*

These are techniques arising from AI, enabling machines to learn, in a more or less autonomous manner, to accomplish tasks without being explicitly programmed.

Machine learning refers to the development, analysis and implementation of methods enabling a machine to evolve through a learning process, and hence fulfill tasks that are difficult or impossible to accomplish by more classical algorithmic means.

There are three big types of machine learning:

– *supervised learning*: the algorithm attempts to predict a phenomenon or a measure based on the history of achievements of the latter. The database is formed of labeled data;

– *unsupervised learning*: does not involve the prediction of a measure; the algorithm rather attempts on its own to detect the characteristic structures or groups in a given set of observations. Data are not labeled; the objective is then to find a relation between data;

– *reinforcement learning*: the intelligent agent observes the effects of its actions and deduces the quality of its actions in order to improve its future actions. The action of the algorithm on the environment generates a return value (reward or punishment) that guides the learning algorithm.

To conclude, the main objective of machine learning is to automatically extract and use the information present in a dataset. But the actual potential of machine learning resides in processing data that were never seen previously, while finding the correct answers. For this reason, the core of machine learning is the amount and quality of data, as well as the choice of the best machine learning algorithm that can be integrated in our data.

3.2.2.4. *Neural networks*

The human brain is composed of a set of interconnected neurons transmitting elaborated models of electrical signals. Dendrites receive the input signals and based on these inputs, a neuron produces an output signal via an axon (Shiffman 2012).

Artificial neural networks draw their inspiration from the biological operation of the human brain, and therefore, by analogy with a biological neuron, an artificial neuron is perceived as an autonomous processor with unidirectional channels for communication with other neurons connected to it.

An artificial neuron has several input channels operating as dendrites, and only one output channel operating as an axon. The connection points between neurons are known as "synapses". The typical operation of an artificial neuron is to calculate a weighted sum of the input signals and generate an output signal if this sum exceeds a certain threshold. The weighted sum of input signals is done by the combination function, which multiplies the input vector by a transformation matrix. The output signal is generated by the function.

Figure 3.2 represents the structure of an artificial neural network.

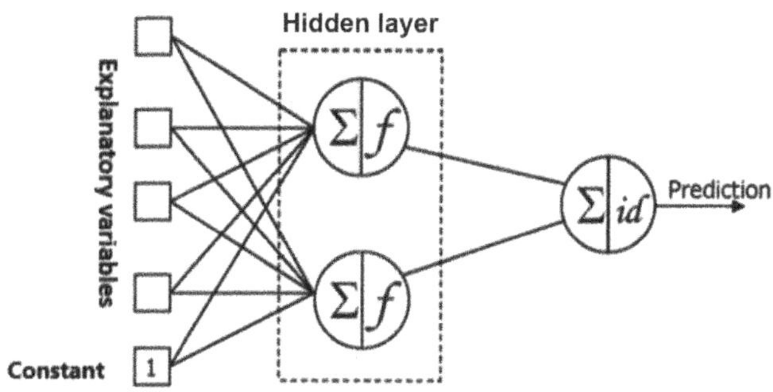

Figure 3.2. *Artificial neural network (Decourt 2018)*

3.2.2.5. *Multiagent systems*

A multiagent system (MAS) is a group of agents, each of which has one or several elementary competences. The purpose is to have these agents work together in order to solve a problem or accomplish a specific task. It is a sort of intelligence distribution, each autonomous agent having only a local view of the problem or an elementary task of the work to be done.

Ferber and Perrot (1995) define an MAS as follows:

"A multi-agent system is composed of the following elements:

– an environment, which is a space generally having a metrics;

– a set of objects located in space; they are passive; they can be perceived, destroyed, created and modified by the agents;

– a set of agents, which are the active entities of the system;

– a set of relations, which interconnect the objects;

– a set of operations enabling agents to perceive, destroy, create, transform and manipulate the objects;

– a set of operators in charge of representing the application of these operations and the reaction of the world to this modification attempt (laws of the universe)".

MASs are generally used when the problem is too complex to be solved by a single system, due to several software or hardware limitations. A specific case of use is when components have multiple interrelations. MAS are an excellent tool to ensure autonomous control in a widely distributed system with very dynamic characteristics.

MASs are certainly the ideal solution for scenarios requiring a system that must dynamically adapt when new components are added or removed and these components must easily adapt when the environment undergoes modifications. It should be kept in mind that one of the most important advantages of MASs is their modularity, which enables simpler programming, in the sense that adding new agents to an MAS poses no significant problem; this explains their scalability (Amraoui 2015).

The interest of the agent-based solution resides in the entire absence of a central entity reacting to agent operation, which provides high resistance and reliability (if an agent breaks down, the system continues to operate).

3.2.2.6. *Internet of Things*

The IoT is a term that generally describes a system in which physical objects are connected to the Internet, now known as "ecosystem of connected objects".

IoT starts in the physical world with sensors that gather information; this information is then forwarded because of system interconnection and integration; data are finally processed and stored in order to be analyzed and used.

An essential characteristic is that IoT can transform ordinary objects into devices. They can be identified by an IP address, record states via sensors and have memory capacity via microchips. The integrated minicomputers enable them to self-control, manage their environment and automatically exchange data. Because of machine learning they are sometimes even capable to recognize and generalize models and draw conclusions in order to adapt to situations and continuously optimize.

3.2.2.7. *Cloud computing*

The official definition of cloud computing was given by Mell *et al.* (2011): "Cloud Computing is a model for enabling convenient, on-demand network access to a shared pool of configurable computing resources (e.g., networks, servers, storage, applications and services) that can be rapidly provisioned and released with minimal management effort or service provider interaction".

Cloud computing stands for providing various hardware and software solutions via Internet. Processor performances, storage space and software environments can be rented by users in order to extend or replace their own infrastructure. Cloud computing gives users the possibility to store an enormous amount of data as well as access it from any place and at any moment. In order to use the stored information, the users must have basic Internet connection.

3.2.2.8. *Big data*

Big data is a term that describes any collection of data whose volume and complexity are such that it is difficult to process it by classical tools for application processing. It is a generic term employed to designate the strategies and technologies used for gathering, organizing and analyzing vast sets of data.

3.3. Network optimization

Network optimization involves improving network operation in terms of security and reliability, performance and rapidity, QoS and, obviously, energy consumption.

This section gives an outline on how AI optimizes networks.

3.3.1. *AI and optimization of network performances*

With the growing demand for Wi-Fi and the integration of connected objects in our lives, the lack of visibility and network control are major factors in computer management. Therefore, continuously ensuring the same QoS and the same performance is a challenge to be met. One promising solution is the implementation of a cognitive network known as intelligent.

Cognitive networks are a specific type of networks capable of learning, predicting and improving performances, security and user experience. These solutions use the cloud, data analysis, machine learning and AI to determine basic performances, follow activity and identify problems.

The large amount of data entering the network from different nodes inevitably requires very high computation power. With the use of AI, all these data must be studied within machine learning. Hence, the network can understand, for example, the moment when the applications reach non-optimal performances or compare a constant flow of historic and current data.

3.3.2. *AI and QoS optimization*

Recent technological progress has enabled the low-cost manufacturing of small elements, such as small wireless sensors used for on-site measurement of ambient conditions. These sensors are used due to their low energy consumption, low radio range, weak memory and low cost.

It is also worth noting that multimedia traffic has recently significantly increased, as recent technologies (IoT, for example) add new traffic types, and especially video flows. According to a Cisco report[1], in 2021 video traffic will be three times that of 2016. Moreover, this traffic represents 82% of the total internet traffic in 2021.

This indicates that the network QoS will continue to be a requirement for the real-time transmission of unpredictable data. Consequently, heavy tasks will still have to be executed in the Cloud.

This being said, FAI have no control and generally no knowledge on the Wi-Fi access points used by mobile devices users, therefore they cannot guarantee that QoS is provided as promised.

The term QoS does not concern only the rate, packet loss, latency or jitter. It is also a matter of availability. To be able to further optimize networks and enable an efficient energy management, high reliability and availability, it is important to provide communications security and this requires AI integration for better dynamic management of the network traffic.

With the use of AI techniques, it is possible to discover the various types of flows being transmitted in the network. Traffic models can thus be obtained, helping in the decision-making process.

In Nowicki and Uhl (2017), the authors consider that multimedia traffic can be more efficiently managed using AI techniques. Their paper proposes an intelligent

1 Cisco Visual Networking Index, Forecast and Methodology, 2016-2021.

system to provide QoS and Quality of Experience (QoE) in the video monitoring of the traffic generated by the equipment in the IoT.

Once lost, network quality cannot be reestablished. But a possibility exists to integrate quality and cognitive intelligence at each end of the connection.

Machine learning can also be useful for mitigating the risks related to network unavailability or to security exploits. For example, with cognitive radio (CR) (Benmammar *et al.* 2012), the application knows that you are preparing to go through a black coverage area and it can consequently be proactive.

Further works exist in the literature and they rely on meta-heuristics, such as the paper by Benmammar (2017). The latter uses the Shuffled Frog Leaping Algorithm (SFLA) meta-heuristics to improve QoS in a CR network. The authors' objective is to maximize throughput and minimize error rate and energy consumption in this type of network.

3.3.3. *AI and security*

According to a study conducted by Cigref (2018), over one in two companies have been hit by cyberattacks, which is enormous in terms of costs.

In France, only 29% of the companies consider cybersecurity a high priority challenge[2]. Even more worrying, only one company in two has implemented a strategy focusing on the fight against cyber risks.

In order to counteract this type of attacks, employees must be aware of basic security measures and of the need to use antivirus software and firewalls. But this is obviously not enough to provide full security and for this purpose there are other more evolved and high-performance solutions based on AI, and more specifically on machine learning. These new methods enable the easier detection of anomalies and generate alerts quite rapidly in order to inform system administrators.

Unlike classical methods such as antivirus software, which make sure the machine contains no specific signature (an indication that a program is considered malicious or not), machine learning-based systems learn how to search for the

2 According to the IPSOS reference document from 2017 and 2018 report by PWC, available at: https://www.pwc.fr/fr/assets/files/pdf/2018/10/ pwc-barometre-cybersecurite-septembre-2018.pdf.

various characteristics of malware in order to learn their behavior and be able to detect them more rapidly; these systems are therefore more flexible.

Techniques known as intelligent are intensely used in the fight against spam and phishing and have yielded good results. The latter can also be used to protect the system against attacks coming from the inside (from ill-intentioned employees). In this sense, AI can conduct a behavioral analysis: it studies the behavior of a computer and makes it possible to alert those in charge with system security in case of deviant computer behavior (data leakage attempt, for example).

This being said, while the operation of intelligent techniques enable them to analyze situations and behaviors, this requires a large amount of data to be able to yield efficient and satisfactory results. A lack of data may therefore lead to false results and hence to false alerts; this is why full automation of a security system is impossible, as human intervention is essential in some cases.

The current AI engines use statistical data for classification (malicious or honest), but their capacity may also become a weakness. Indeed, the machine learning engine is equivalent to humans in terms of learning capacity, but at a larger scale and with much higher speed.

For an AI to be efficient, its learning engine is fed a very large amount of information. As it receives information, the engine builds a statistical model that enables it to autonomously determine when the sought-for phenomenon occurs.

In fact, since the strength of statistical algorithms lies in their capacity to recognize models and diagrams, the attackers may progressively adapt their behaviors to appear normal or act in a manner that induces confusion.

Moreover, many systems may detect anomalies at the beginning, but after a while they learn to accept them as normal behaviors. This offers the attackers an advantage, as they can mask their activities by observing normal behaviors, such as the use of the "https" protocol to send data to a server.

Attackers may add execution stages that do not contribute to reaching their objective, but are designed to make the process appear normal. Moreover, weak signals that seem harmless to the human analyst may prove efficient in deceiving machine learning algorithms.

Finally, it can be said the AI brings an additional security layer and it can significantly slow down computer hackers. Indeed, even though a human can deal with many threats per hour, he can be rapidly overtaken by a significant flow of threats. AI is helping humans in their processing of security incidents and can even rapidly suggest or apply remediation actions.

Intelligent systems are frugal consumers of CPU and RAM resources compared to traditional antivirus software and do not necessarily need an Internet connection. They no longer need to know the threat in order to block it, nor do they require permanent updates, as the model relies on a statistical approach. The system analyzes a large amount of data having the various characteristics of a file, its potential signature, size and code, all these recurrent series of bits. Then it is assigned a score, which tells if a file may be executed or not.

3.3.4. *AI and energy consumption*

When used in energy production or consumption, AI operates by means of sensors installed in the control systems. This enables real-time processing of data. Because of this, system anomalies or malfunctions are detected and dealt with much more rapidly. Once the problems are highlighted, the faulty system or equipment can be replaced; this enables maximum optimization of energy efficiency.

The development of connected objects associated with the use of AI technologies enables the deployment of tools aiding the intelligent consumption and management of energy. It also enables the deployment of systems for real-time prediction and management based on storage and self-consumption.

Energy consumption and production prediction for real-time energy management purposes can be done using regression algorithms such as random forest or restricted Boltzmann machine. In terms of energy efficiency improvement, k-means clustering methods can be used.

3.4. Network application of AI

3.4.1. *ESs and networks*

3.4.1.1. *ES for machine maintenance*

An expert diagnostic system is a series of computer applications that integrate a large base of knowledge or expert reasoning and automatically infer the root causes of the observed anomalies.

It is a computer system intended to determine what causes an equipment failure by analyzing and representing the knowledge and reasoning of one or several maintenance specialists. It makes it possible for an operator with average or even poor technical knowledge to look for the probable cause of a failure, by providing the system with minimum information, such as the type of defective machine and the observed failure mode.

Knowledge modeling is the most important part in the design of a computer-aided diagnosis system, therefore the elements to be studied and their relations should be properly defined.

The facts base and the work space of the computer-aided diagnosis system are enriched as the system is deployed in terms of selecting the most probable cause of the failure and its adequate remedy.

To build a base of valid reasoning, diagnosis expert knowledge can be formalized as rules, decision trees, propositional logic, etc.

In the case of ES aiding machine maintenance (Kaushik *et al.* 2011; Raja'a and Jassim 2014), the knowledge base contains the machine-specific knowledge provided by maintenance experts. This knowledge takes the form of facts and rules.

In this case, learning relies on the data generated by the equipment of the network to be diagnosed and it essentially involves interpolation or induction-based solving of the reverse problem of fault and alarm propagation in a telecommunication network.

3.4.1.2. *ES for the diagnosis of a multiplexer network*

In Lor (1993), the authors developed a system for the diagnosis of multiplexer networks. Diagnostic expert knowledge is classified into two categories: generic expert knowledge and precise diagnostic task-specialized expert knowledge. The ES uses a database of static and dynamic information required during the diagnosis process. This information refers to the relations between logical entities (channel groups) and the physical entities (equipment and links), such as routing information, attributes of physical entities and incidence relations between network nodes and links.

A line can be diagnosed in two stages. The first step is to collect the available data on the line, such as the power levels transmitted and received by the equipment of the line, supply voltages, polarization currents, temperatures of the line equipment, transmission error counters and alarms observed. Each datum is stored in a key performances identifier (KPI). Then the predefined expert rules use these KPI to generate an indication or a final diagnostic decision referred to as "conclusion".

Diagnostic expert de la ligne cliente - DELC (Expert Diagnosis of the Client Line) is a rule-based ES developed by Orange Labs France for the automated diagnosis of the Digital Subscriber Line (xDSL) and of the Gigabit Passive Optical Network (GPON) of Fiber To The Home (FTTH) type.

3.4.2. *CBR and telecommunications networks*

It often happens, particularly in a complex CBR system, such as the one involved in the diagnosis of a telecommunications network with a broad diversity of anomaly signatures, that an adaptation of preexisting solutions is required.

A CBR system for fault diagnosis in a DUMBO network was proposed in Melchiors and Tarouco (1999). This system uses knowledge on diagnostic cases stored in an incident ticket system in order to propose diagnostic solutions to new anomalies. This system aims to facilitate the stages of diagnosis and resolution of network management problems.

The knowledge unit of a CBR system is the case and not the rule. A case is easier to articulate, examine and evaluate than a rule (Hounkonnou 2013). A CBR system is also capable of learning from its own errors/failures and improving its performances. The phase of evaluation of solutions to new problems is worth paying attention to. Indeed, poor evaluation may drive the integration of erroneous cases in the knowledge base and thus cause the entire system to drift.

3.4.3. *Automated learning and telecommunications networks*

Machine learning can be employed to diagnose a larger number of faults than rule-based ESs are able to; it can diagnose problems outside of its expertise field, although in such cases its performances are lower.

The diagnosis of a telecommunications network requires a comprehension of the phenomenon of fault and alarm propagation in this network. This comprehension enables the acquisition of relevant knowledge in order to automatically solve the reverse problem of fault and alarm propagation.

To be able to diagnose anomalies occurring in a telecommunications network, the diagnosis system may be a learning system, which has the induction capacities enabling it to use its knowledge base to find the root causes of new anomalies, previously unknown to it.

This method no longer uses a reasoning base specialized on accurate diagnosis tasks, such as ES and CBR, but knowledge on the behavior or operation of a telecommunications network. This knowledge is used to build a structured and explicit representation of the network operation. The complexity involved by the development of a model-based diagnosis system is due to the fact that a large-scale telecommunications network is very often heterogeneous and dynamic, with a large number of equipment of various types.

In Łgorzata Steinder and Sethi (2004), the authors explain that model building is only the first stage in the network diagnosis based on a model of the respective network. The second stage involves the development or implementation of an algorithm based on the model. The algorithm starts with the entities that triggered the alarms and explores the relations between the network entities formalized by the model. Their algorithm is also able to determine the correlated alarms and thus localize the offending entities of the network.

In Yu *et al.* (2009) and Fan *et al.* (2012), the authors explain how automated learning, and especially artificial neural networks, can be used for intrusion detection.

The model-based approach is easy to deploy and modify and it is appropriate for a large-scale network if the information related to network resources is available.

3.4.4. *Big data and telecommunications networks*

3.4.4.1. *Big data and customer service improvement*

Telecommunications companies collect enormous amounts of data from call recordings, mobile phone use, network equipment, server logs, invoicing and social networks, thus providing much information on their customers and their network. With big data technologies, telecommunications companies will use these data to improve their activity through advanced analyses.

With the rapid expansion of smartphones and other connected mobile devices, providers of communication services must rapidly process, store and draw information from the diversified volume of data going through their networks. Big data analyses are used in order to:

– help improve efficiency by optimizing network use, improving customer experience and strengthening security;

– predict the periods of intensive network use and target the stages to reduce congestion;

– identify the customers with the highest chances of failing and target the stages enabling turnover prevention;

– identify the customers with the highest chances of having difficulties paying their invoices and target the stages for payment collection improvement.

Due to the significant volume of data, it is important to process them near the source, and then efficiently transfer them to various data centers for further use.

Real-time analysis of events is key to timely analysis of network services in order to improve customer satisfaction. Abandoned calls, locations with average network coverage quality, low download speed, unacceptable waiting time, etc., are examples of potential analysis subjects.

In network applications, the key to successful use of big data is to focus on problems and not on data points.

In terms of network administration, big data are collected from probes deployed in various points, as well as by means of network layer software installed on client and server equipment. When they are presented in a system infrastructure with standard administration, a part of this information may correspond to common management practices.

3.4.4.2. *Big data and security*

In big data, there are more significant volumes of data, but these are especially exponential, variable and from different sources.

Since the company can have a view on entire volumes of data carried daily by its information system, instead of waiting for problems to occur in order to process them, it can attempt to identify all the potentially signaling events.

Once the risk is identified, a protection is implemented to prevent its propagation. A proactive view on computer security can then be deduced, all the more so as the accuracy of the collected information enables better identification of threats by tracing them back to their source.

3.4.5. *MASs and telecommunications networks*

Agent techniques have as their main task developing knowledge engineering that reduces information processing to knowledge-based reasoning. These techniques also enable the development of software engineering techniques adapted to service delivery.

The telecommunications field offers the perspective of open environments, either in the Web or in future network services. Moreover, it enables the exploration of various agent techniques: mobile agents, web assistants and knowledge reasoning agents.

Several decades ago, when companies wanted to have a private telecommunications network, they used a telecommunications infrastructure of their own. Then these requests were satisfied by private networks composed of connections let to an operator. The use of these connections can be subcontracted to the operator. These connections, which are not part of the public networks, ensure a QoS, for example the band required between several given points and full confidentiality of exchanged data.

Virtual private networks (VPNs) are offered as private networks implemented on public networks. With a VPN, any increasing demand for temporary connections can be met and the bandwidth that is not used by a company at a given instant is potentially available for another use. MAS can be used for the automation of VPN supply requiring several network services providers and for the automation of network resources negotiation in this context.

The agents designate software components for the decentralized control or monitoring of network resources. They are used for the development of cooperation strategies enabling the coordination of the assignment or supervision of resources depending on various authorities, as well as for the development of strategies for the control of network overload that could be generated by the signaling related to new services.

3.4.5.1. *MAS and CR*

In Mir (2011), the author proposes cooperation between PUs (Primary Users) and SUs (Secondary Users), and among SUs only. Agents are deployed on user terminals to cooperate and agree on contracts governing the spectrum assignment. SU agents coexist and cooperate with PU agents in a CR environment *ad hoc* using messages and decision-making mechanisms. Given that the internal behaviors of the agents are cooperative and selfless, this enables them to maximize the utility function of other agents without additional costs in terms of exchanged messages.

Nevertheless, resource allocation is an important challenge in CR systems. It can be realized by negotiation between secondary users (Li 2009; Qian *et al.* 2011). In Qian (2011), the authors propose an agent-based model for spectrum negotiation in a CR network. In this model, instead of direct spectrum negotiation between PU and SU, a broker agent is included. This means that the PU or SU equipment does not require high intelligence, since it does not need to conduct spectrum detection or other more complicated tasks of the CR. The objective of this negotiation is to maximize the benefits and profits of the agents in order to satisfy the SU. The authors proposed two situations, the first of which uses only one agent that uses and dominates the network, while in the second several agents are competing.

One study was conducted by Xie *et al.* (2007) on the CR in the wireless local area network (WLAN) referring to the possibility of introducing agent technology; in other terms, they try to solve the problem of radio resource allocation by associating WLAN resources management in a decentralized environment by means of MAS. For this purpose, they propose an agent-based approach for information sharing and decision distribution among multiple WLAN in a distributed manner.

In Amraoui (2015), a multi-agent architecture is proposed involving three levels: the first one is the physical level, where the authors made several remarks on the type of terminal used, followed by the cognitive level, where they proposed a MAS-based modified cognition cycle, and finally the behavioral level, where they studied various potential behaviors of the agents during spectrum negotiation.

3.4.5.2. *MAS and transport networks*

Transposing agent-oriented notions to the transport field is in agreement with the characteristics of these two fields. Autonomy, distributed behaviors and partially observable environment characteristics are in fact present. The existing approaches focus on the properties of MASs: emergence, self-organization and cooperation. Moreover, the evolution of a large number of vehicles on a shared road network perfectly corresponds to the resource conflict problems studied via MAS (Guériau 2016).

Intelligent transport systems offer a set of tools relying on the latest advances in terms of computation power, communication and perception, in order to produce a supervised, integrated, universal and approachable system. MAS seem to be the best path to improvement, both in terms of interaction and distributed computation.

Intersection management can also be improved by the use of MAS. Indeed, the agents can be deployed at the intersection level or for each light, and their cooperation enables the optimization of cycles in response to the demand.

MAS can also prove practical for traffic control and congestion management, when agents can cooperate and negotiate to provide better road traffic management and a more intelligent transport network.

3.4.6. *IoT and networks*

In order to be able to adapt in real time to a given situation, connected devices must understand the value of the information they gather and learn from each other. Thanks to AI and analytics, they then make adequate decisions within autonomous systems.

Nowadays, the objects we use in our homes, offices, hospitals and factories are in the connection phase. Providing them with autonomous learning and personalization capacities, AI will bring them in a disruption phase.

The combination of IoT and AI leads to an actual change of perspective. Sectors such as security, health, industry and energy are potential beneficiaries of the advantages offered by this combination.

An example of application of the AI + IoT combination in the field of security is that of security software embedded in connected cameras featuring computer vision, which are able to identify a person in a crowd and alert the competent authorities because of shape recognition techniques. In particular, the software can count the number of persons in a room, identify a person with a criminal record or as a wanted person, authorize a person's access in confidential areas, etc.

In the health field, the connected objects combined with AI have a place in most situations. For example, due to a connected camera or a connected pair of glasses, they enable the identification of disease symptoms. The connected sensors can transmit vital data concerning a patient to a platform featuring AI, which calls a nurse or a care assistant if needed and is able to provide all the details of the event.

In Srinidhi *et al.* (2018), the authors propose several AI-based algorithms for IoT network optimization. Indeed, they approach several types of algorithms, such as genetic algorithms, which use multiobjective criteria for the selection of the best sensors with a maximal storage space.

3.5. Conclusion

With the acceleration and transformation of telecommunication networks because of new technologies, operators must improve the efficiency of their services while reducing costs.

In our opinion, the use of AI and of data science can improve network performances, reliability and security.

Indeed, with AI, the network is able to automatically react to any potentially occurring significant overload. A network will be able to detect an overload and automatically create a number of virtual machines required for the amount of input traffic.

Fault diagnosis in a large-scale telecommunications network is a complex problem that presents an interest for both telecommunications operators and the AI community. This problem is the object of many research works and various approaches were proposed relying on ESs, CBR systems and machine learning-based systems.

In our view, in the future telecommunications networks will become fully autonomous and will no longer depend on human intervention, due to AI and especially to big data technologies.

3.6. References

Amraoui, A. (2015). Vers une architecture multiagents pour la radio cognitive opportuniste. PhD Thesis, University of Tlemcen, Algeria.

Benmammar, B. (2017). Optimisation de la QoS dans un réseau de radio cognitive en utilisant la métaheuristique SFLA (Shuffled Frog Leaping Algorithm) [Online]. arXiv:1703.07565.

Benmammar, B., Amraoui, A., and Baghli, W. (2012). Performance improvement of wireless link reliability in the context of cognitive radio. *International Journal of Computer Science and Network Security*, 12(1), 15–22.

Cigref (2018). Cybersécurité: visualiser, comprendre, décider [Online]. Report. Available at: https://www.cigref.fr/wp/wp-content/uploads/2018/10/Cigref-Rapport-Cybersecurite-Visualiser-Comprendre-Decider-Octobre-2018.pdf.

Decourt, O. (2018). Les réseaux de neurones expliqués à ma fille [Online]. Available at: https://od-datamining.com/knwbase/les-reseaux-de-neurones-expliques-a-ma-fille/.

Fan, W., Bouguila, N., and Ziou, D. (2012). Variational learning for finite Dirichlet mixture models and applications. *IEEE Transactions on Neural Networks and Learning Systems*. IEEE, 23(5), 762–774.

Ferber, J. and Perrot, J.-F. (1995). *Les systèmes multiagents: vers une intelligence collective*. InterEditions, Paris.

Guériau, M. (2016). Systèmes multiagents, auto-organisation et contrôle par apprentissage constructiviste pour la modélisation et la régulation dans les systèmes coopératifs de trafic. PhD Thesis, University Claude Bernard Lyon 1.

Hounkonnou, C. (2013). Active self-diagnosis in telecommunication networks. PhD Thesis, European University of Brittany and University of Rennes 1.

Kaushik, A., Barnela, M., Khanna, S., and Kumar, H. (2011). A Novel Expert System for PC Network Troubleshooting and Maintenance. *International Journal of Advanced Research in Computer Science (IJARCS)*, 2(3), 201–203.

Łgorzata Steinder, M. and Sethi, A.S. (2004). A survey of fault localization techniques in computer networks. *Science of Computer Programming*, 53(2), 165–194.

LI, Husheng (2009). Multi-agent Q-learning of channel selection in multi-user cognitive radio systems: A two by two case. *2009 IEEE International Conference on Systems, Man and Cybernetics*. San Antonio, Texas, USA, 1893–1898.

Liang, Q., Feng, Y., Lin, G., Xiaoying, G., Tian, C., Xiaohua, T., Xinbing, W. and Mohsen, G. (2011). Spectrum trading in cognitive radio networks: an agent-based model under demand uncertainty. *IEEE Transactions on Communications*, 59(11). IEEE, 3192–3203.

Lor, K.W.E. (1993). A network diagnostic expert system for Acculink multiplexers based on a general network diagnostic scheme. *Proceedings of the 3rd IFIP/IEEE International Symposium on Integrated Network Management*. San Franciso, USA.

Melchiors, C. and Tarouco, L.M.R. (1999). Fault management in computer networks using case-based reasoning: DUMBO system. *International Conference on Case-Based Reasoning*. Springer, Berlin, Heidelberg, 510–524.

Mell, P. and Tim, G. (2011). The NIST definition of cloud computing [Online]. Available at: https://nvlpubs.nist.gov/nistpubs/Legacy/SP/nistspecialpublication800-145.pdf.

Mir, U. (2011). Utilization of cooperative multiagent systems for spectrum sharing in cognitive radio networks. PhD Thesis, University of Technology of Troyes.

Nowicki, K. and Uhl, T. (2017). QoS/QoE in the heterogeneous Internet of Things (IoT). In *Beyond the Internet of Things*, Batalla, J.M., Mastorakis, G., Mavromoustakis, C.X., and Pallis, E. (eds). Springer, Basel, 165–196.

Raja'a, A.K. and Jassim, R.O. (2014). Expert system to troubleshoot the wireless connection problems. *International Journal of Science, Engineering and Computer Technology*, 4(8), 238.

Shiffman, D. (2012). *The Nature of Code: Simulating Natural Systems with Processing.* Shannon Fry, USA.

Srinidhi, N.N., Dilip Kumar, S.M., and Venugopal, K.R. (2018). Network optimizations in the Internet of Things: A review. *Engineering Science and Technology*, 22(1), 1–21.

Xie, J., Howitt, I., and Raja, A. (2007). Cognitive radio resource management using multiagent systems. *4th IEEE Consumer Communications and Networking Conference.* Las Vegas, USA.

Yu, W., He, H., and Zhang, N. (2009). Advances in neural networks. *6th International Symposium on Neural Networks* (ISNN 2009). 26–29 May 2009. Wuhan, China.

4

Multicriteria Optimization Methods for Network Selection in a Heterogeneous Environment

Fayssal BENDAOUD
ESI-SBA, Sidi Bel Abbès, Algeria

4.1. Introduction

As a consequence of its boom during the last decade, the telecommunications sector plays an important and essential role in our present-day societies. The most significant innovation in this field is undoubtedly "wireless" technology. The connection possibility offered by Hertzian waves gave rise to several types of wireless networks. These networks compete for user attention, offering higher performances, especially in terms of quality of service (QoS). From the perspective of TCP/IP model, the IP and link layers are the most important. Consequently, IP must cover the key functionalities associated with wireless networks, particularly the management of handover and of GPS localization and mobility. Moreover, due to the broad range of choices of radio access technologies proposed by many providers of cellular/wireless networks, and considering the QoS requirements of modern applications, choosing a network type proves to be a complicated task. A single and unified approach to network choice is therefore required, and this is referred to as network selection in a heterogeneous environment. A heterogeneous wireless environment is a set of wireless equipment using various radio access technologies. In a heterogeneous environment, the user is not connected to a single network, but

Intelligent Network Management and Control,
coordinated by Badr BENMAMMAR.

he/she can rather benefit from all the networks present in his/her surroundings. Compared to the traditional homogeneous wireless networks, such an environment has several advantages:

– better QoS, since at each instant an application launches an algorithm for network choice and connects the user to this network;

– service availability: if a radio access technology disappears, the application connects the user to another network;

– load balancing of the networks and efficient use of the spectrum: at each instant, the application places the user in the least loaded network, expecting him/her to have good QoS.

No single radio access technology is able to provide these advantages. Nevertheless, a heterogeneous wireless environment has many problems that remain to be solved such as:

– network interoperability and network selection;

– mobility management;

– handover;

– QoS/experience;

– network interference.

This chapter focuses on network selection in a heterogeneous environment and QoS. The main objective of a heterogeneous environment is to materialize the “Always Best Connected” concept, offering mobile users the possibility to benefit from networks with different performances. Nowadays, there is a broad range of radio access technologies, WLAN (essentially IEEE802.11) and cellular networks (UMTS, HSPA and LTE). This variety constitutes the heterogeneous environment. The network selection procedure aims to select the best network among the available ones. The decision is however very difficult and complex, due to the large number of parameters involved in the selection process, such as the cost of using such a network, QoS and energy consumption. Therefore, the basic definition of the network selection problem is the dynamic and automatic choice of the best wireless access network, taking into account the already mentioned parameters. In classical cellular systems, the mobile user has a contract-based connection with a single network, so there is no network selection. According to the new view on networks, the user can choose at any moment the ideal network for him/her and use it to establish his/her session, having no contractual obligation to stay connected to a single network (Figure 4.1).

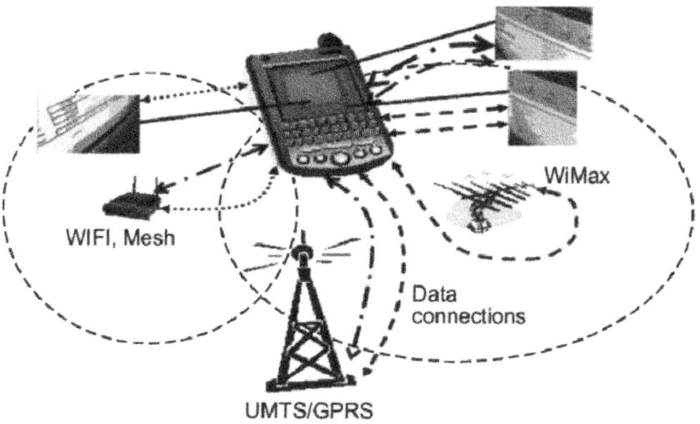

Figure 4.1. *A heterogeneous environment (Bendaoud 2018)*

The chapter is organized as follows: section 4.2 presents multicriteria optimization methods; section 4.3 proposes a modification of one of the multicriteria optimization methods (accompanied by a choice motivation and a comparison between the modified method and the unmodified one); and section 4.4 is dedicated to a conclusion opening on a range of perspectives.

4.2. Multicriteria optimization and network selection

Prior to the third-generation networks, the notion of network selection did not even exist, as the user had a contract-based connection with a single network. Afterwards, the rapid development of network technologies led to an impressive increase in services and applications that use the Internet and, at the same time, the number of users has increased. Nowadays, the users who have intelligent peripheral equipment are trying to acquire the ABC concept. Obviously, no single radio access technology can provide the ABC concept; a change of ideas is therefore essential, more specifically a passage from homogeneous systems (users who have a contract-based connection with a single network) to heterogeneous systems. The objective of a heterogeneous environment is to materialize the ABC concept by offering mobile users the possibility to benefit from networks with various performances.

Network selection in a heterogeneous environment can be compared to multicriteria optimization problems due to the number of parameters involved and the criteria to be met. This mathematical approach has been widely used to solve the network selection problem (Kovvali *et al.* 2015; Scherzer and Scherzer 2015; Wu *et al.* 2015; Bendaoud *et al.* 2018b). Other methods such as fuzzy logic and game theory have been proposed to deal with this problem (Watanabe *et al.* 2008; Alkhawlani and Ayesh 2008). The focus of this section is on the use of multicriteria optimization methods for the selection of the best network.

4.2.1. *Network selection process*

As already said, the network selection process involves a switch between radio access technologies aimed at always having the best service at any moment. Hence, when a user with a multimode device discovers the existence of several networks in his area, he/she should be able to select the best network for him/her among the existing networks and use it during his/her session. The various networks offer several characteristics in terms of delay, jitter, throughput and packet loss rate (PLR). This is why nowadays it is essential to correctly select the best network for system efficiency (Figure 4.2).

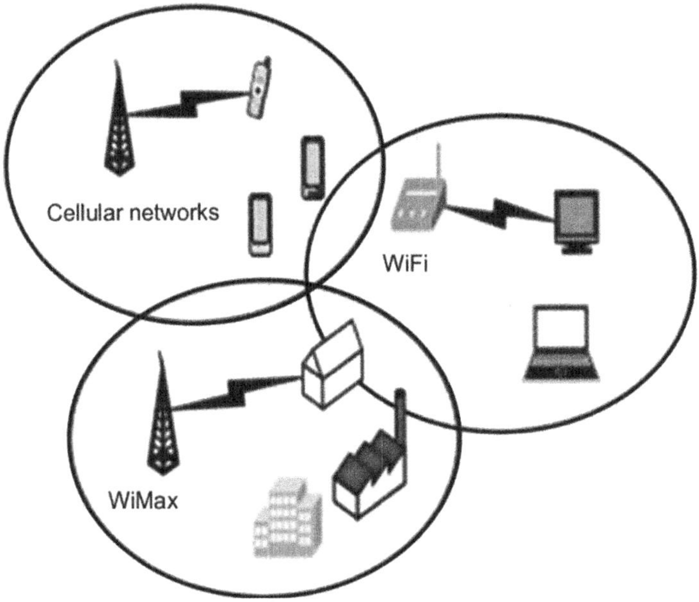

Figure 4.2. *Heterogeneous environment (Bendaoud 2018)*

The network selection procedure is the general case of the handover; this latter can be centralized, meaning that it is done by the operator, or decentralized, which means that it is user-centered.

In the network-centered approach, the operator controls the entire process and makes decisions. The users obey these decisions and implement them. This can be considered a correct strategy for preventing problems such as the egoistic behavior of users who always try to get the best network at the same time, which leads to a congestion of the loaded network. Furthermore, this approach assumes a single operator and several networks; it cannot be used for several operators. In the user-centered approach, users make the decisions themselves; this approach is decentralized and can easily generate congestion, due to users' egoistic nature. Nowadays, almost all operators offer 3G and 4G radio access, besides Wi-Fi connections; The first (centralized) approach is therefore preferable.

The process of choosing the best network depends on many parameters, such as battery status, energy needed for the required services, received signal strength (RSS), the network use cost, allocated bandwidth and user preferences (excellent quality or good quality/price ratio). These parameters are classified into various classes and sets (Figure 4.3).

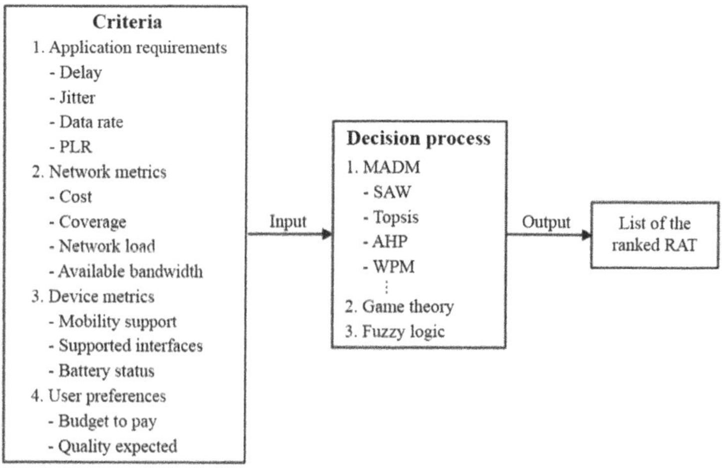

Figure 4.3. *Network selection process (Bendaoud* et al. *2018b)*

The details on the parameters involved in the selection process are the following:

– parameters related to network conditions: these group information on network conditions, such as network load, coverage area (radius), time to establish network connection, available bandwidth, etc.;

– parameters related to application requirements: these correspond to the information on the threshold needed for the application to be in a normal state.

These include, among others, required throughput, delay for packet transmission, jitter, PLR and energy required by the application. These are QoS parameters;

– parameters related to user preferences: these represent the information needed by the application users; they include, among others, the budget that the user is ready to pay and the preference for a certain cost/QoS ratio;

– parameters of the mobile equipment: these group information on the user device, such as battery status at instant t and mobility management.

Considering all these parameters involved and their rankings, network selection is indeed a very complex task. There are several works in the literature dealing with network selection; these studies focus on the optimization of network selection decision for users, so that applications benefit of the best possible QoS at instant t and users maintain their connection to this best network as long as possible. Several authors dealt with the best network selection problem, trying to find the most appropriate solution. The next section presents multicriteria optimization methods, which are the best known and most commonly used approach to this problem.

4.2.2. *Multicriteria optimization methods for network selection*

Multicriteria optimization is a branch of mathematics dealing specifically with optimization problems with several objective functions, criteria and alternatives, while the objectives to be optimized relate to a single problem. Everyone uses this approach in everyday life, for example when buying a car that has several characteristics. In the decision matrix, the lines represent criteria and the columns are alternatives. The values a_{ij} describe the performances of alternatives A_j in relation with criteria C_i. In the decision, weights $w_1, \dots, w_m$ are assigned to criteria, and the weighting w_i reflects the relative importance of criterion C_i in the decision process. The criteria weights are generally determined on a subjective basis: they represent the opinion of one or more experts. On the other hand, there are objective methods enabling the determination of criteria weights. They will be dealt with in the following section of the chapter. This approach is very well adapted to the network selection problem due to the multicriteria nature of the network (Sgora *et al.* 2010). The basis of this approach is the following:

– alternatives: the set of players to be ranked. For network selection, the alternatives are the lists of networks;

– attributes: parameters and criteria used in the decision-making process. For network selection, parameters include throughput, jitter and delay;

– weighting: this reflects the importance of a parameter in the decision process.

Several methods were proposed in this context, such as simple additive weight (SAW), technique for order preference by similarity to ideal solution (TOPSIS), weighted product model (WPM) and analytical hierarchy process (AHP) (Salih *et al.* 2015). SAW, TOPSIS and WPM methods rank only the alternatives; therefore they require other methods for criteria weighting. On the other hand, AHP is both a ranking and a weighting method; it contains a process for criteria weighting. Table 4.1 summarizes some of the best-known weighting methods.

Method	Description
Entropy	$w_j = 1 - \frac{1}{N} * \sum_{i=1}^{N} x_{ij} * \ln x_{ij}$
Variance	$w_j = \frac{1}{N} * \sum_{i=1}^{N} x_{ij}$
Eigenvector	$(B - \lambda \times I) \times w = 0$

Table 4.1. *Weighting methods*

4.2.2.1. *Simple additive weight*

The SAW method, also known as "mathematical methods for multicriteria analysis", is a simple multicriteria decision technique, and one of the most commonly used, based on the weighted average score. This method assumes the processed data have the same unit. Consequently, to be homogeneous, data should be normalized for each parameter (Savitha and Chandrasekar 2011a; Abdullah and Adawiyah 2014). Finally, the alternative with the highest or lowest value (depending on the objective function) is selected. The mathematical formula for SAW is:

$$R_{SAW} = \sum_{i=1}^{n} w_j * r_{ij} \qquad [4.1]$$

where:

– R_{SAW}: value of each alternative;

– w_j : weighting value of criterion *j*;

– r_{ij} : normalized value of criterion *j* and alternative *i*.

Once the process is completed, the maximal/minimal value is chosen depending on the tendency of the objective function (maximization or minimization).

Since its creation, SAW has been used in the context of the network selection problem (Nguyen-Vuong *et al.* 2008; Savitha and Chandrasekar 2011a). The method is used as network selection and ranking mechanism in Salih *et al.* (2015), Nguyen-Vuong *et al.* (2008), and Savitha and Chandrasekar (2011b); it is used with game theory in order to rank networks, as the authors understood that SAW used as a single method cannot yield the best performances in all cases. The problems related to multicriteria optimization methods, and in particular to the SAW method, will be revisited at the end of section.

4.2.2.2. *Technique for order preference by similarity to ideal solution*

The technique for order preference by similarity with the ideal solution (TOPSIS) is a multicriteria decision analysis method; it was originally proposed by Hwang and Yoon in 1981, and then developed by Yoon in 1987 and by Hwang *et al.* in 1993. TOPSIS is an aggregation and compensation method based on the concept according to which the chosen solution must be at the shortest geometrical distance from the positive ideal solution and at the longest geometrical distance from the negative ideal solution (Olson 2004). The TOPSIS process is conducted as follows:

– an evaluation matrix composed of m alternatives and n criteria is created; the intersection of alternatives and criteria is represented in the form x_{ij}, which gives a $(x_{ij})_{m*n}$ matrix;

– $(x_{ij})_{m*n}$ matrix is normalized in order to obtain the values $(r_{ij})_{m*n}$ by using one of the normalization methods;

– the weighted normalized decision matrix is calculated with:

$$t_{ij} = w_j * (r_{ij})_{m*n}$$

$$w_j = \frac{W_j}{\sum_{k=1}^{n} W_k} \text{ with } \sum_{i=1}^{n} w_i \textit{ and } W_k \text{ being the criteria weight vector}$$

– the best and poorest alternatives are determined:

$$A_b = \{b_j^p = \max t_{ij} \,|j \in J_+\}$$

$$A_{wo} = \{wo_j^N = \min t_{ij} \,|j \in J_-\}$$

with J_+ and J_- being two sets containing criteria with positive and negative impact, respectively;

– the separation measure for each alternative is calculated:

$$D_p = \sqrt{\sum_{i=1}^{n} (w_j^2 * (r_{ij} - b_j^p)^2)}$$

$$D_N = \sqrt{\sum_{i=1}^{n} (w_j^2 * (r_{ij} - wo_j^N)^2)}$$

– the ideal relative solution is calculated:

$$R_{TOPSIS} = D_p D_p + D_N$$

TOPSIS was used in several works in the literature (notably in Bakmaz *et al.* 2007; Sgora *et al.* 2010; Savitha and Chandrasekar 2011a). In the latter article, the authors compare the performances of vertical handover with two other methods, TOPSIS and SAW; they noted that TOPSIS is better than SAW. In Bakmaz *et al.* (2007), the authors select the network via the TOPSIS method and they conclude that TOPSIS is sensitive to user preferences and to the values of input parameters.

4.2.2.3. *Weighted product model*

The WPM method, also known as "multiplicative exponential weighting method" is similar to SAW (Lahby *et al.* 2012). The difference is that the addition in the SAW method is replaced by multiplication in WPM, besides the fact that each alternative is compared to other decision criteria. Each ratio is raised to a power equivalent to the relative weight of the corresponding criterion. The mathematical description of this method is as follows:

$$R_{WPM} = \prod_{i=1}^{n} (r_{ij})^{w_j} \quad [4.2]$$

Depending on the choice of r_{ij} , there are two variants of WPM:

$$r_{ij} = x_{ij} \; ou \; r_{ij} = \frac{x_{ik}}{x_{jk}} \quad [4.3]$$

Savitha and Chandrasekar (2011b) compared SAW and WPM in the context of vertical handover: they use relative standard deviation as a metric for comparing the two methods and conclude that WPM is better than SAW. In TalebiFard and Leung (2011), the authors use the WPM method for network selection; they conclude that the WPM method is a more robust approach for dynamic decision making.

4.2.2.4. *Analytical hierarchy process and gray relational analysis*

AHP involves the decomposition of a complicated problem into several simple subproblems. AHP includes the following stages:

– problem decomposition into hierarchical subproblems, where the higher node is the final objective and alternatives are listed for each criterion;

– pair-to-pair comparison of attributes and their transformation into numerical values from 1 to 9;

– calculation of weights (weighting) for each hierarchical level;

– weight synthesis and global weight.

As for the gray relational analysis (GRA) method, it is used to rank the candidate networks. It relies on the following principle:

– data normalization takes into account three situations: the highest is the best, the lowest is the worst and the nominal (moderate) is the one required;

– the definition of the ideal sequence in the three situations considered: the ideal sequence contains the higher limit, the lower limit and the moderate limit;

– calculation of the relational coefficient (GRC): the most favorable sequence is the one with the highest GRC.

The AHP method is generally associated with GRA; AHP is often used for weighting (weight of alternatives), while GRA is used for the ranking of its alternatives. Shen *et al.* (2010) used a modified version of AHP and compared it with the traditional AHP using a quality of experience (QoE) criterion; finally, their numerical results show that the proposed modified AHP is better than the conventional AHP, which results from proper load balancing for networks. In Lin and Hsu (2003), the authors use AHP in order to rank various criteria for online advertising networks; the proposed work provides an objective and efficient decision model that advertising agencies can use when choosing an Internet advertising network.

4.2.2.5. *Discussion*

In summary, multicriteria optimization methods are widely used for solving the network selection problem, as the latter has the same diagram and the same characteristics as the problems solved by this type of methods. Moreover, these methods are known for their convenient use, clarity and low calculation complexity. Nevertheless, they have some drawbacks as follows:

– these methods do not have the same performances for different services (Voice over IP [VoIP], video calls and the best effort service). For example, one method can generate good performances with VoIP service and poor performances for video services, which is not ideal;

– rank reversal is a problem of these methods; it is a phenomenon occurring in multicriteria optimization methods when an exact replica or copy of an alternative is added or eliminated. Wang and Luo (2009) showed that rank reversal is frequent in most known multicriteria optimization methods. This problem was approached in other works (Huszak and Imre 2010; Shin *et al.* 2013) proposing modifications of these methods, but the original versions of the methods are plagued by the rank reversal phenomenon;

– AHP method is very complicated and requires complex calculation involved in weighting vector calculation (eigenvalues calculation).

Due to these reasons, multicriteria optimization methods can be considered a good solution, but the absence of a method that compensates the previously mentioned aspects is a problem.

Advantages	Drawbacks
Easy to understand	Rank reversal phenomena
Easy to implement	High complexity for certain methods such as AHP and ELECREE
Very good results in certain cases	Good performance with certain applications and poor results for others

Table 4.2. *Advantages and drawbacks of multicriteria optimization methods*

4.3. "Modified-SAW" for network selection in a heterogeneous environment

This section presents a method known as "modified-SAW" (Bendaoud *et al.* 2017) to overcome the problems of the existing solutions to the network selection problem. Indeed, the existing multicriteria optimization methods all have the well-known problem of rank reversal when an alternative is added or deleted; these methods have further problems, such as sensitivity to user preference (notably in TOPSIS method) and alternatives degradation with poor values for certain WPM attributes. The proposed method does not have the above-mentioned problems and, at the same time, it outperforms other methods.

It is worth noting that, from the perspective of new generation networks, heterogeneous wireless access is a promising feature, as the users have the capacity "to be flexible enough" in the selection of the most appropriate network according to their needs. Therefore, network selection is an important task for the proper operation of the entire heterogeneous communication system. Indeed, network selection process involves the failover between Radio Access Technology (RAT) aiming to provide the user with the best network. Hence, when a user with a multimode device discovers the existence of several RAT in the same area, he/she should be able to select the best network to obtain the desired service (Figure 4.2).

4.3.1. *"Modified-SAW" proposed method*

The proposed network selection process, "modified-SAW" (Bendaoud *et al.* 2017), aims to provide the user with good QoS during the call session and, at the same time, ensure a correct allocation of users to each network. The proposed solution enables the user to have access to the best network among the networks available at current instant, ensuring that the selected network is accessible, which means not overloaded. This process is repeated several times until the end of the user's call session.

Hence, when a user wants to use a particular service, he/she sends a request to the operator. This request contains information such as the required service and the user's battery level. The other parameters required for network selection are collected by the operator. Then the latter initiates the ranking of available networks. Results are then sent to the user, who selects the best available network. He/she will obviously choose the network with the best rank, since an overloaded network cannot normally be ranked as best, simply because it has a longer delay and a lower throughput; its performance is therefore poor in terms of QoS (Figure 4.4).

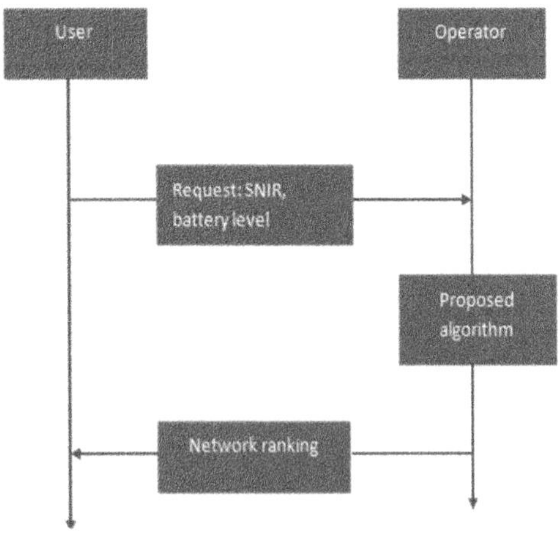

Figure 4.4. *Network selection process*

In our description based on Figure 4.4, we have two agents: the mobile user, who is looking for the best RAT, and the operator, who initiates the ranking of the available networks. The objective function relies on a simple idea. The system is formulated as a minimization function in which the lowest value for each criterion gives the highest order for the network. Consequently, the network has the highest local gain. This process is repeated until all the criteria are evaluated. Its representation is as follows. Consider that for each network "*i*", the following calculation is made:

$$R_i = \sum_{j=1}^{m} income_{ij} \quad [4.4]$$

$$income_{ij} = (\alpha - k_{ij}) * w[j] \quad [4.5]$$

$$k_{ij} = \min(Vect_{ij}) \quad [4.6]$$

where:

– α: fixed integer number, equal to the number of alternatives;

– k_{ij}: ranking order of network *i* for criterion *j*;

– $Vect_{ij}$: column vector of matrix mat where *j* is fixed;

– *w[j]*: weighting vector associated with criterion "*j*" for variant "*i*";

– *i*: alternative;

– *j*: criterion;

– mat [n] [m]: input matrix, represented by Table 4.2.

Let us start by dividing the input matrix in a set of column vectors in order to obtain a group of vectors equal to the number of criteria. For each vector, the networks (alternatives) are ranked depending on their data values. Then each network receives a local income equal to the mathematical multiplication of "α", minus the corresponding network rank value, and the weighting value of the criterion. "α" is a fixed value equal to the number of criteria. This process is repeated for the other criteria. The global income value is equal to the sum of all local incomes for each network (see equation [4.4]). Let us use the weight value of each criterion in order to differentiate the criteria, which is what all multicriteria optimization methods do. It is well known that delay and PLR criteria are the most important. They represent the most important criteria of QoS.

```
Data: The matrix mat [m][n] and the weight vector
            W[], α=m
Result: Income vector for each network
for (j=0; j<m; j++) do
    for (i=0;i<n;i++) do
            tab[i]=mat[i][j]
    end
    tabind=Sort(tab)
    for (k=0;k<tabind.length;k++) do
            income[tabind[k]] = income[tabind[k]]+(α − k)*w[k]
    end
    return income[]
end
```

Algorithm 4.1. *Modified SAW method*

The local income system based on network rank for each criterion enables the income control for each alternative depending on its rank for all the criteria. For example, a network proposing a delay value of 30 ms and another one proposing a value of 100 ms do not have the same local revenue.

In this example, a small delay value is good and provides the network with a higher gain. On the other hand, a higher bandwidth materializes in higher local income. The criteria evaluation process is repeated until all criteria are covered.

The best case for each network "*i*" is to have a minimal value for criterion "*j*", hence the highest local ranking, which means $k_{ij} = 0$, and the income is $R_{ij} = \alpha * w[j]$, knowing that in this study α = 5 (number of criteria).

The least favorable case occurs when the network has the maximal value for criterion "j", which means that $k_{ij} = \alpha - 1$, hence the income becomes $R_{ij} = w[j]$.

The use of the weighting concept gives more meaning to the objective function, which ranks the networks depending on their values for each criterion. The weighting vector enables a distinction between significant criteria and less important ones, depending on application demands and attributes, which leads to more significant income for networks with the highest values for important criteria.

```
Data: vector tab[]
Result: Sort function
int t[][], double min;
int ind, cpt, deb=0;
int a, nb, val=0:
while (deb<t.length) do
        while (tab[a]<0) do
                a++;
        end
        min=tab[a]; ind=a;
for(int i=0;i<tab.length;i++) do
        if (min>tab[i] and tab[i]>=0) then
                        min=tab[i]; ind=i; nb=0;
        end
        else if (min==tab[i] and a!=i) then
                        nb++
        end
end
for(int p=deb; p<deb+nb+1;p++) do
        for(int h=0;h<2;h++) do
                if(h==0) then
                                t[p][h]=val;
                end
                else
                                t[p][h]=ind+cpt; cpt++;
                end
        end
end
```

Algorithm 4.2. *Sorting function of M-SAW algorithm*

In this study, the weight vector (weighting) is used in association with value "α". This modification has mainly two advantages:

– it prevents a situation in which a network having a good value for an unimportant criterion gets the same income as another network with a good value for an important criterion. This situation occurs in the SAW method because of the use of the ordinary weight vector (without the modification proposed in this work);

– the weighting concept designates the representation of the requirements of the applications in the system. This is how applications are distinguished, since each of them has specific requirements. VoIP requires a minimum delay to be respected; for the video service, the PLR should not exceed a certain threshold. This information is transformed into numerical values in order to obtain the weighting vector; this is done with the eigenvalues method.

This work takes into consideration many parameters in the input matrix, such as cost, energy consumption, average throughput reached, average delay, PLR and network. Table 4.3 presents an example of such a matrix; the values were obtained in simulations on NS3.

	Bandwidth	Delay	Lost packets	Cost	Energy
Wi-Fi	1–11	100–150	0.2–8	1	–
3G	1–14	25–50	0.2–8	5	–
LTE	1–100	60–100	0.2–8	2	–

Table 4.3. *Values of simulations*

The energy consumption parameter is related to the battery of the mobile peripheral equipment, and also to the session length of the application, therefore with a high battery level the energy consumption is not that significant in the system, because the mobile equipment is able to sustain an uninterrupted session. On the other hand, if the loading level of the battery is low and the session duration is long, this means that the session can be interrupted because of battery. Therefore, if the battery is weak, this parameter is very important and has a more significant weight in the system in order to prevent battery depletion, and therefore a session interruption. The value of energy consumption is defined by the following equation (Huang *et al.* 2012):

$$P[mJ/s] = \alpha_u * th_u + \alpha_d * th_d + \beta \quad [4.7]$$

where:

– α_u, α_d and β : parameters with values that differ from one RAT to another (Huang *et al.* 2012);

– th_u and th_d: throughput of the uplink and throughput of the downlink.

The power expressed in mJ/s reflects the dependence of energy on the user session length.

4.3.2. *Performance evaluation*

This section focuses on evaluating the proposed model performances and comparing them with above-described multicriteria optimization methods based on the input data in Table 4.3. Three types of services are therefore considered: VoIP, video service and best effort service. These services are represented in the system by the associated weighting vectors, since each type of application has specific QoS requirements. Many users think that the problem is solved by more bandwidth (throughput), but simply having more bandwidth is not the ideal solution for all applications. In packet switched networks, QoS is influenced by various factors, since packets may be subjected to many alterations when traveling from source to destination. The factors that determine QoS are bandwidth, packet reception delay, PLR and variation of packet reception time, known as "jitter".

Several important technical factors are as follows:

– low throughput: due to the large number of users sharing the same network resources, the effective throughput that can be provided to a data flow may be too weak and insufficient for real-time multimedia services (Bendaoud 2018);

– latency: the time for a packet to reach destination may be very long, as it can be blocked in long queues or take a different path to avoid congestion (Bendaoud 2018). In certain cases, excessive latency may render an application such as VoIP or online games inoperable. Therefore, latency can be considered a very important parameter for the ideal behavior of the network;

– jitter: it is the variation of time required for the packets to travel from their sources to their destinations. Packet reception delay varies depending on its position in the router queue and on the pathway between source and destination (Bendaoud 2018). This delay variation can seriously affect the quality of audio and video streaming;

– lost packets: routers may fail to deliver (abandon) certain packets if these data packets are corrupted, or if the packets arrive when router buffer memories are already full (Bendaoud 2018). The receiver application may ask for the retransmission of this information (packets), which may lead to serious delays in the overall transmission.

These are the most important factors that influence an application in a wireless network. Therefore, the weighting vector is used for the representation of these requirements in the system. Performance evaluation involves two parts:

– in the first part, our proposal is compared with existing multicriteria optimization methods, and in the normal case when no RAT disappears during the selection process;

– in the second part, a network disappears from the list of available networks; it is the proof that multicriteria optimization methods are plagued by a rank reversal phenomenon and that this problem does not arise in the method that we propose.

Table 4.4 represents the input matrix used in our comparisons; this matrix relies on the values presented in Table 4.3.

	Bandwidth	Delay	Lost packets	Cost	Energy
N(0)	1.730	105.85	7.94	0.2	1.00
N(1)	5.076	134.88	6.70	0.2	2.6
N(2)	6.849	43.98	2.84	1	6.26
N(3)	6.329	32.15	3.05	1	5.86
N(4)	66.66	95.15	6.32	0.4	12.78
N(5)	62.5	99.73	5.80	0.4	10.28

Table 4.4. *Input matrix*

The weighting values for each type of VoIP application, video service and *best effort* service are generated with the eigenvector method (see equation [4.8]). The eigenvector method was chosen because it was already used in AHP method. Therefore, for the sake of consistency and correctness, we decided to use the same method in order to have the same weighting vectors for all the methods:

$$(\text{mat} - \lambda \times \text{I}) \times \text{w} = 0 \qquad [4.8]$$

where:

– mat: input matrix;

– λ: eigenvalue;

– w: associated eigenvector containing the weighting values;

– I: identity matrix.

Table 4.5 presents the weighting vector for each type of application.

	Bandwidth	Delay	Lost packets	Cost	Energy
VoIP	0.047	0.486	0.371	0.047	0.047
Video	0.458	0.101	0.302	0.074	0.063
Best effort	0.299	0.146	0.146	0.108	0.299

Table 4.5. *Weighting vectors*

Let us start with the first part of this study, the ordinary case, when all networks are available.

4.3.2.1. *Simulation 1: availability of all networks*

In this case, all networks are available; this means that the networks located in the user coverage area do not disappear. In this situation, we consider the already presented network selection algorithms and compare them with the proposed M-SAW algorithm. The simulations concern three types of applications, namely VoIP, video service and best effort services.

VoIP

The first case of this simulation concerns the VoIP application. The latter is represented via a weighting vector that focuses on delays and packet loss. For VoIP, throughput is not very significant, as packets are small and can be transmitted with relatively low throughput (Lewis and Pickavanc 2006). Table 4.6 summarizes the results related to the first case, which is VoIP. The analysis in Table 4.6 relies on Table 4.4 as data input.

Method	Ranking
SAW	N(1) N(0) N(2) N(3) N(5) N(4)
TOPSIS	N(3) N(2) N(0) N(1) N(5) N(4)
WPM	N(3) N(2) N(0) N(1) N(5) N(4)
AHP	N(3) N(2) N(5) N(4) N(1) N(0)
M-SAW	N(3) N(2) N(4) N(5) N(0) N(1)

Table 4.6. *Ranking for VoIP*

This simulation focuses on the total order of networks, and not only on the network ranked best, as the top network may rapidly become loaded after a given lapse of time, then unavailable, namely overloaded. This is why it is essential to have an overall optimal ranking of networks. As the results in Table 4.6 indicate, TOPSIS, WPM and M-SAW methods give the same order for the first two networks N(0) and N(1). For the third position, our method M-SAW chooses N(4), but TOPSIS and WPM choose N(0), while AHP chooses N(5). Let us now compare the performances of N(4), N(0) and N(5) to see what method made the proper choice.

N(0) has a delay of 105.85 and a PLR of 7.94. N(4) proposes a delay of 95.15 and a PLR of 6.32.

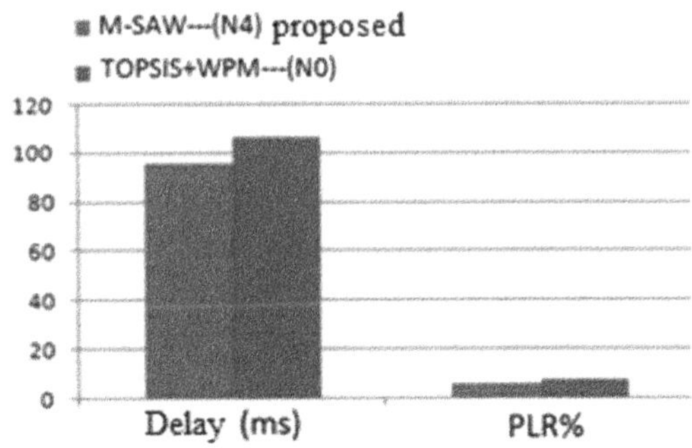

Figure 4.5. *Delay and lost packets comparison for N(0) and N(4). For a color version of this figure, see www.iste.co.uk/benmammar/network.zip*

Therefore, N(4) is better than N(0) and this shows that the performances of our method are higher than those of TOPSIS, WPM and AHP (Figures 4.5 and 4.6). For VoIP, we proved that the proposed M-SAW method gives the best ranking order compared to multicriteria optimization methods.

Video service

The second case of our simulations is dedicated to video applications. The associated weight (weighting) vector places greater importance on throughput than on transmission delays and packet losses. Throughput is more important, because of the large data volumes that must be transmitted without loss (Szigeti *et al.* 2013).

Method	Ranking
SAW	N(1) N(0) N(2) N(3) N(5) N(4)
TOPSIS	N(5) N(4) N(2) N(3) N(5) N(0)
WPM	N(4) N(5) N(3) N(2) N(1) N(0)
AHP	N(2) N(3) N(5) N(4) N(1) N(0)
M-SAW	N(4) N(2) N(3) N(5) N(1) N(0)

Table 4.7. *Ranking for video service*

In Table 4.7, M-SAW and WPM select N(4) and TOPSIS chooses N(5). N(4) has high throughput and small delay. N(5) has better PLR. But in this case of "video service", importance is placed on throughput and delay. Therefore, the best choice is N(4) (Figure 4.6).

Moreover, for the second order of the ranking, WPM method selects N(5) and M-SAW selects N(2). N(5) has a bandwidth of 62.5, a delay of 99.73 ms and a PLR of 5.80. N(2) proposes a throughput of 6.85, a delay of 43.98 ms and a PLR of 2.84.

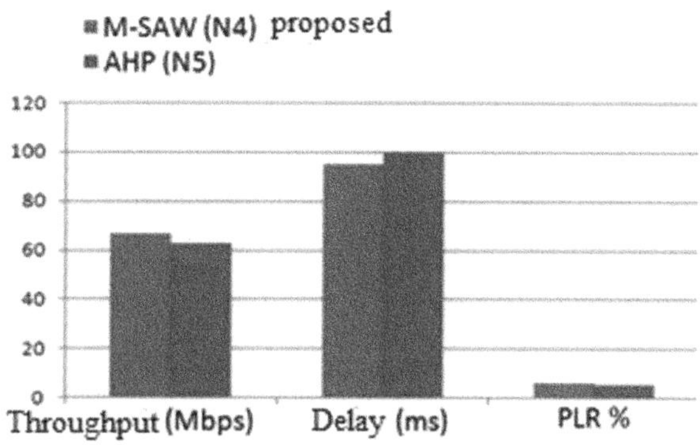

Figure 4.6. *Throughput and delay comparison for N(5) and N(4). For a color version of this figure, see www.iste.co.uk/benmammar/network.zip*

Based on these values, it can be seen that N(5) proposes a higher throughput, but that of N(2) is also good and can amply support the interactive service (video service).

The property used in this case is that of providing the user with the minimal value meeting the application requirements. In other words, if application requirements are met, the network is judged as acceptable and the user can choose it.

For the other parameters (delay and lost packets), N(2) is very good compared to N(5) (Figure 4.7). It can be noted here that the throughput parameter has the monopoly over the ranking decision in TOPSIS, which means that the network having the best throughput forces the algorithm to neglect enormous delay and high PLR.

Therefore, according to Figure 4.7, our M-SAW method yields the best choices for this type of service (video service).

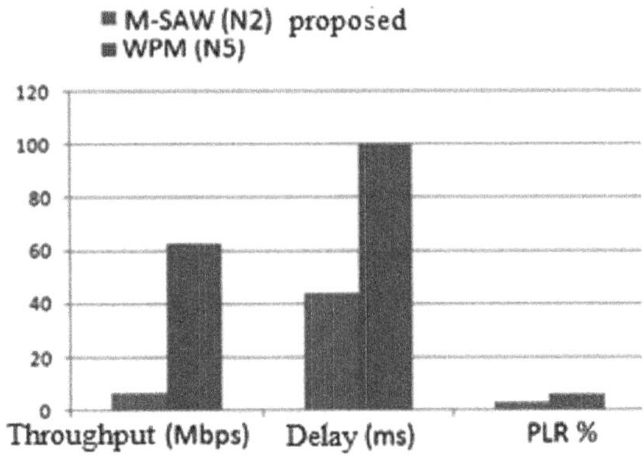

Figure 4.7. *Comparison between N(2) and N(4). For a color version of this figure, see www.iste.co.uk/benmammar/network.zip*

Best effort services

The third type of application employed relates to best effort applications, such as mail service and web browsing. The associated weight vector places no particular importance on any criteria: it gives equal value to all criteria, expecting to obtain high throughput. Packets are transmitted and the user ignores the time required for their transmission and the quality of received packets.

Table 4.8 summarizes the results of the third case (file download). WPM and M-SAW have the same orders for the first and second ranks; for the third position, WPM chooses N(1) and M-SAW chooses N(2). As Figure 4.8 clearly shows, N(2) is better than N(1).

Method	Ranking
SAW	N(1) N(0) N(2) N(3) N(5) N(4)
TOPSIS	N(5) N(4) N(2) N(3) N(1) N(0))
WPM	N(4) N(5) N(1) N(2) N(3) N(0)
AHP	N(5) N(1) N(4) N(2) N(3) N(0)
M-SAW	N(4) N(5) N(2) N(3) N(1) N(0)

Table 4.8. *Ranking for best effort service*

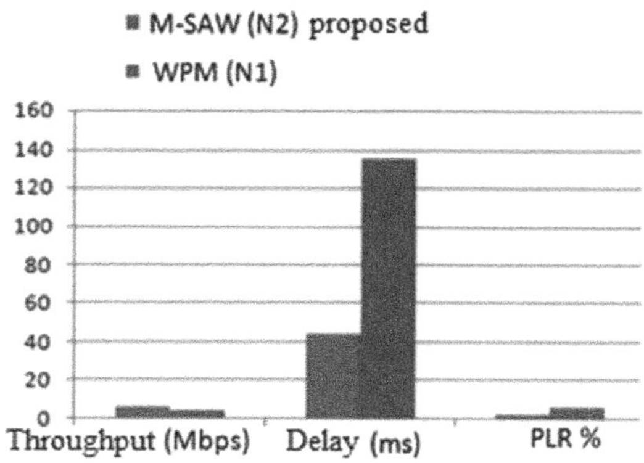

Figure 4.8. *Comparison between N(2) and N(1). For a color version of this figure, see www.iste.co.uk/benmammar/network.zip*

In this initial part of the simulation, a comparative study is conducted for M-SAW algorithm and the multicriteria optimization algorithms presented in the first sections. Three cases were simulated: VoIP, video service and best effort services. In all three cases, the proposed algorithm yields more accurate results and gives each time the total rank adapted for the system. The figures presented in this section indicate that our algorithm gives in all cases the total exact order of all the alternatives.

4.3.2.2. *Simulation 2: rank reversal*

This section deals with the rank reversal problem, also known as "ranking anomaly". There are two objectives to be met in this second simulation: the first is to confirm that this problem (rank reversal) occurs in multicriteria optimization methods when one line of the matrix is deleted (one alternative disappears); the second objective is to show that our proposed M-SAW method does not have this problem. If the second objective is reached, it can be stated that our proposed method is very well adapted to multicriteria problems in general and to network selection in particular.

Let us now assume that one network disappears; this situation can be used as a case study for rank reversal phenomenon. Therefore, one network is eliminated (one alternative of the input matrix), for example network N(4) in Table 4.4, and selection algorithms are applied to the remaining networks in the case of VoIP service (the same procedure is applicable to the other types of services). Table 4.9 summarizes the results.

Method	Ranking
SAW	N(1) N(3) N(2) N(0) N(5)
TOPSIS	N(5) N(3) N(2) N(0) N(1)
WPM	N(5) N(3) N(2) N(0) N(1)
AHP	N(0) N(1) N(5) N(3) N(2)
M-SAW	N(3) N(2) N(5) N(0) N(1)

Table 4.9. *Ranking when one alternative disappears*

The results in Table 4.9 indicate two things: first, all multicriteria optimization methods (TOPSIS, SAW, WPM and AHP) are exposed to rank reversal phenomenon, and this result confirms what the authors state in Savitha and Chandrasekar (2011b), namely that all multicriteria optimization methods are subjected to rank reversal. This confirms that the direct use of this type of method is not the proper choice for network selection.

Second, this (rank reversal) problem is not present in our M-SAW method when one network disappears. According to this result, it can be said that M-SAW is better adapted for network resolution than multicriteria optimization methods. Furthermore, M-SAW is not complex (easy to understand and implement and rapid in terms of run time), as only a few modifications were brought to the SAW method, which is known for its low complexity.

4.3.2.3. *Discussion*

In summary, multicriteria optimization methods are used to obtain the ranking order of alternatives in a multiple criteria problem; this ranking order is not always optimal. One method may rank a network as best, and the user would want to connect to that network; given the number of users that select it as the best network, he/she discovers that it is loaded. Therefore, the automated selection of the best network easily loads this best network and after a while the latter no longer accepts the connection requests from users. This is why it is important to have a correct and optimal list of rankings of available networks. Consequently, if the best network is loaded, the user goes to the next best network; hence, the ranking algorithm must optimally and accurately order the networks. At the same time, the algorithm must prevent rank reversal, given that a loaded network becomes inaccessible and is deleted from the list of candidate networks. Traditional multicriteria optimization methods do not meet any of these demands, which are satisfied by M-SAW method. The selection process is conducted on the operator side to take advantage of the processing capacity operator and the permanent (energy) supply, which provides both efficiency and rapidity; moreover, the operator has all the information concerning networks and users. Therefore, the algorithm seeks to find the optimal total ranking of networks and not only the best network in a list of networks. Having the optimal order of networks enables the user to be connected at each instant to the best available network among existing networks. A second advantage of this algorithm is that it operates properly in the normal case and also when a network disappears. In the latter case, traditional multicriteria methods are confronted with rank reversal problems.

4.4. Conclusion

To be able to find at each instant the best network, the idea was to classify existing networks and obtain the optimal ranking, while the operator provides the users with the best network available in the list of ranked networks. This work presents a method named "M-SAW", in which the objective function relies on the relative order of each alternative for each criterion at each process iteration. This method enabled us to obtain a "glutton" algorithm, which yields good results. Indeed, as simulations prove, our method surpasses the methods referred to by the literature and which were already used. The tests conducted indicated that all the traditional multicriteria optimization methods are subjected to rank reversal phenomenon. Our algorithm does not have this problem and remains consistent by providing the same ranking order while eliminating the network that disappeared.

4.5. References

Abdullah, L. and Adawiyah, C.W. (2014). Simple additive weighting methods of multi criteria decision making and applications: A decade review. *International Journal of Information Processing and Management (IJIPM)*. 5(1), 39–49.

Alkhawlani, M. and Ayesh, A. (2008). Access network selection based on fuzzy logic and genetic algorithms. *Advances in Artificial Intelligence*, 8(1), 1.

Bakmaz, B., Bojkovic, Z., and Bakmaz, M. (2007). Network selection algorithm for heterogeneous wireless environment. *IEEE 18th International Symposium on Personal, Indoor and Mobile Radio Communications*. 3–7 September 2007, Athens, Greece, 1–4.

Bendaoud, F. (2018). Management of joint radio resources in heterogeneous networks Beyond 3G. PhD Thesis, École supérieure en informatique, Sidi Bel Abbès, Algeria.

Bendaoud, F., Abdennebi, M., and Didi, F. (2018). Network selection in wireless heterogeneous networks: A survey. *Journal of Telecommunications and Information Technology*, 4, 64.

Bendaoud, F., Didi, F., and Abdennebi, M. (2017). A modified-SAW for network selection in heterogeneous wireless networks. *ECTI Transactions on Electrical Engineering, Electronics, and Communications*, 15(2), 8–17.

Huang, J., Qian, F., Gerber, A., Mao, Z.M., Sen, S., and Spatscheck, O. (2012). A close examination of performance and power characteristics of 4G LTE networks. *Proceedings of the 10th International Conference on Mobile Systems, Applications, and Services*. Low Wood Bay, United Kingdom.

Huszak, A. and Imre, S. (2010). Eliminating rank reversal phenomenon in GRA-based network selection method. *2010 IEEE International Conference on Communications*. 23–27 May 2010, Cape Town, South Africa.

Kovvali, S.K., Boyle, C., Ghai, R., Amand, J.S., and Alden, G. (2015). U.S. Patent No. 9,001,682. U.S. Patent and Trademark Office.

Lahby, M., Cherkaoui, L., and Adib, A. (2012). An intelligent network selection strategy based on MADM methods in heterogeneous networks. *International Journal on Wireless & Mobile Networks*, 4(1), 83–96.

Lewis, C.S. and Pickavanc, S. (2006). *Selecting MPLS VPN Services*. Cisco Press, Indianapolis.

Lin, C.T. and Hsu, P.F. (2003). Adopting an analytic hierarchy process to select Internet advertising networks. *Marketing Intelligence & Planning*, 21(3), 183–191.

Nguyen-Vuong, Q.T., Ghamri-Doudane, Y., and Agoulmine, N. (2008). On utility models for access network selection in wireless heterogeneous networks. *IEEE Network Operations and Management Symposium*. 7–11 April 2008, Salvador Da Bahia, Brazil, 144–151.

Olson, D.L. (2004). Comparison of weights in TOPSIS models. *Mathematical and Computer Modelling*, 40(7–8), 721–727.

Salih, Y.K., See, O.H., Ibrahim, R.W., Yussof, S., and Iqbal, A. (2015). A user-centric game selection model based on user preferences for the selection of the best heterogeneous wireless network. *Annales des télécommunications*, 70(5–6), 239–248.

Savitha, K. and Chandrasekar, C. (2011a). Trusted network selection using SAW and TOPSIS algorithms for heterogeneous wireless networks. *International Journal of Computer Applications*, 26, 8.

Savitha, K. and Chandrasekar, C. (2011b). Vertical handover decision schemes using SAW and WPM for network selection in heterogeneous wireless networks. *Global Journal of Computer Science and Technology*. 18–24.

Scherzer, S., and Scherzer, T. (2015). Method and system for selecting a wireless network for offloading. U.S. Patent No. 9,148,843. U.S. Patent and Trademark Office.

Sgora, A., Vergados, D.D., and Chatzimisios, P. (2010). An access network selection algorithm for heterogeneous wireless environments. *The IEEE Symposium on Computers and Communications*, IEEE publisher. 22–25 Jun 2010. Riccione, Italy, 890–892.

Shen, D.M., Tian, H., and Sun, L. (2010). The QoE-oriented heterogeneous network selection based on fuzzy AHP methodology. *The Fourth International Conference on Mobile Ubiquitous Computing, Systems, Services and Technologies.* October 25–30, 2010. Florence, Italy, 275–280.

Shin, Y.B., Lee, S., Chun, S.G., and Chung, D. (2013). A critical review of popular multi-criteria decision making methodologies. *Issues in Information Systems*, 14(1), 358–365.

Szigeti, T., Hattingh, C., Barton, R., and Briley Jr, K. (2013). *End-to-End QoS Network Design: Quality of Service for Rich-Media & Cloud Networks.* Cisco Press, Indianapolis.

TalebiFard, P. and Leung, V.C. (2011). Context-aware mobility management in heterogeneous network environments. *JoWUA*, 2(2), 19–32.

Wang, Y.M. and Luo, Y. (2009). On rank reversal in decision analysis. *Mathematical and Computer Modelling*, 49(5–6), 1221–1229.

Watanabe, E.H., Menasché, D.S., de Souza e Silva, E., and Leao, R.M.M. (2008). Modeling resource sharing dynamics of VoIP users over a WLAN using a game-theoretic approach. *The 27th Conference on Computer Communications*. IEEE, Phoenix, 915–923.

Wu, J., Cheng, B., Yuen, C., Shang, Y., and Chen, J. (2015). Distortion-aware concurrent multipath transfer for mobile video streaming in heterogeneous wireless networks. *IEEE Transactions on Mobile Computing*, 14(4), 688–701.

PART 3

AI and the Cloud Approach

5

Selection of Cloud Computing Services: Contribution of Intelligent Methods

Ahmed Khalid Yassine SETTOUTI
Abou Bekr Belkaid University, Tlemcen, Algeria

5.1. Introduction

Nowadays, each electronic device generates data related to its use, environment or state. Once generated, data are and will always be important for their generator and/or user.

The Web boom was immediately followed by a sudden increase in the use of computer devices. As a direct consequence, there is a growing amount of data generated by these devices. This rapid expansion gave rise to new needs and requirements in terms of resources for the management of generated information. This is how cloud computing emerged.

Later, the number of cloud computing providers continued to grow. Consequently, choosing one of them became a complex task, especially when unconventional requirements are involved. Furthermore, each provider has its own fees, quality measures, advantages and drawbacks. Therefore, choosing a provider or service became even more difficult.

Intelligent Network Management and Control,
coordinated by Badr BENMAMMAR.

This chapter is a state of the art focusing on the selection, composition, evaluation or recommendation of cloud computing services. The objective, approach or method of the surveyed works must be related to artificial intelligence (AI). The literature features many attempts such as Ahmed *et al.* (2019), who prefers to focus on the selection of cloud computing services in the case of heterogeneous cloud federation, while Whaiduzzaman *et al.* (2014) aims at approaches using multiple choice-based multidecision methods. A similar approach is adopted by Le *et al.* (2014) limiting the set of surveyed works to those using fuzzy logic. Finally, Papathanasiou *et al.* (2015) resumed the work of Whaiduzzaman *et al.* (2014) and added an analytical comparison.

This chapter aims to give the readers an overview on the current works dedicated to cloud computing services that use AI. The objective is to save bibliographic research time for researchers in the field and to facilitate introduction for researchers in need of clarification.

The rest of this chapter is structured as follows:

– section 5.2 presents several general notions required for the comprehension of this chapter;

– section 5.3 lists several similar works in the field of cloud computing services selection;

– section 5.4 presents the monitored works, submits them to critical analysis and offers several improvement suggestions;

– section 5.5 reviews the research advances in the selection of cloud computing services using AI and proposes work that could be conducted in the near future.

5.2. Scientific and technical prerequisites

This section describes several scientific or technical notions required for the comprehension of this chapter. First, a simple, but quite thorough definition of cloud computing is provided. At the end, the concept of AI is presented, including an explanation of what it is and what it is not.

5.2.1. *Cloud computing*

This section briefly presents the concept of cloud computing (definition, characteristics, deployment models and levels of services).

Cloud computing is a model enabling access to remotely configurable (physical or virtual) resources (Mell and Grance 2011).

On the one hand, access must be ubiquitous and on customers demand. On the other hand, it should require minimal customer efforts and no interaction with the service provider (Mell and Grance 2011).

The computerized cloud is expected to provide customers with a certain quality, such as a minimal availability rate (generally 99.95%) (Hayes 2008; Mell and Grance 2011).

5.2.1.1. *Cloud computing characteristics*

Although the previous definition (see section 5.2.1) lists a set of characteristics that distinguish cloud computing from other models and architectures (such as virtualization, on-demand computing and ubiquitous computing), it is important to independently describe each criterion. The following segments provide details on cloud computing properties.

5.2.1.1.1. On-demand service

The customer can allocate as many resources as needed. They can do this at any moment with no service provider intervention (Ahmad *et al.* 2017).

5.2.1.1.2. Access via Internet

All the resources of the cloud are accessible through the Internet (Armbrust *et al.* 2010). Whether heavy (desk applications) or light (web applications), they are available to any Internet user (Mell and Grance 2011).

5.2.1.1.3. Resource grouping

Various resources are grouped for a set of customers (Syntec Numérique 2012). Depending on the processes load and number, resources are allocated and released (Hayes 2008).

Consumers have no knowledge on the exact location of their processing (Mell and Grance 2011). However, they can choose the datacenter where their services are processed, but not a specific server (Armbrust *et al.* 2010).

5.2.1.1.4. Rapid elasticity

Resources can be provisioned, deleted, expanded or reduced as easily and rapidly as possible (Mell and Grance 2011). The customers' perception is that resources are unlimited, accessible and modifiable at any moment and from any connected location, as if they were local (Hayes 2008).

5.2.1.1.5. Measured service

Users pay only for what they use, so the service is measured by its use (Syntec Numérique 2012). This may vary depending on the type of service, but the principle is the same. For example, a storage service (Google Drive, which is free up to 10 GB and available after payment beyond that) can measure the allocated space, without billing the other resources such as the processor and the bandwidth, while a database service (Heroku, for example, which is free for a given database size, but chargeable larger sizes) can measure the size of tables, without including the cost of their creation and/or processing.

5.2.1.2. *Deployment models*

There are four different ways for deploying a computerized cloud. The main differentiation criterion is the targeted audience.

5.2.1.2.1. Public cloud

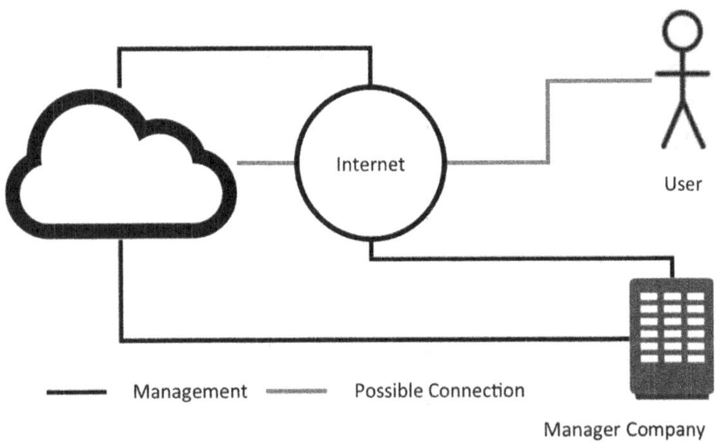

Figure 5.1. *Public deployment of cloud computing. For a color version of this figure, see www.iste.co.uk/benmammar/network.zip*

Although managed by a private company (such as Microsoft, which manages its Azure service for the general public), the cloud computing model for the general public can be accessed by any connected Internet user (Mell and Grance 2011). Consequently, though the least secured, it is not the least used (Hayes 2008).

Figure 5.1 gives an illustration of cloud computing model for the general public. A public cloud service is always accessible. The only condition to be met by the customer is being connected to the Internet.

As with any equipment, it has to be managed by an organization, a company, etc., but this can be done locally or via the Internet.

5.2.1.2.2. Private cloud

Cloud infrastructure or part of the cloud is dedicated to a specific company (Ahmad *et al.* 2017). An Internet user who is not related to the company cannot access equipment considered *private* (Mell and Grance 2011). However, the infrastructure can be managed by the respective company (customer) or by a third party (manager) (Syntec Numérique 2012).

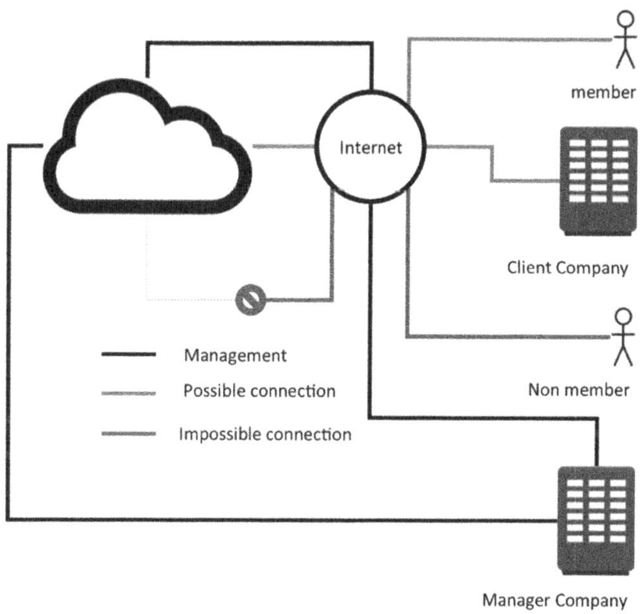

Figure 5.2. *Private deployment of cloud computing. For a color version of this figure, see www.iste.co.uk/benmammar/network.zip*

Figure 5.2 illustrates an example of private deployment of cloud computing. A member of the client company can access dedicated equipment, while another cannot. On the other hand, the equipment can be managed by a third organization (managing company) or by the company itself (client).

5.2.1.2.3. Community cloud

The (entire or partial) cloud infrastructure is dedicated to a community of companies sharing the same requirements and preferences specified in their *Service Level Agreements* (Mell and Grance 2011).

The respective infrastructure can be managed by the whole community, a part of it, a member company or a third party organization (Syntec Numérique 2012).

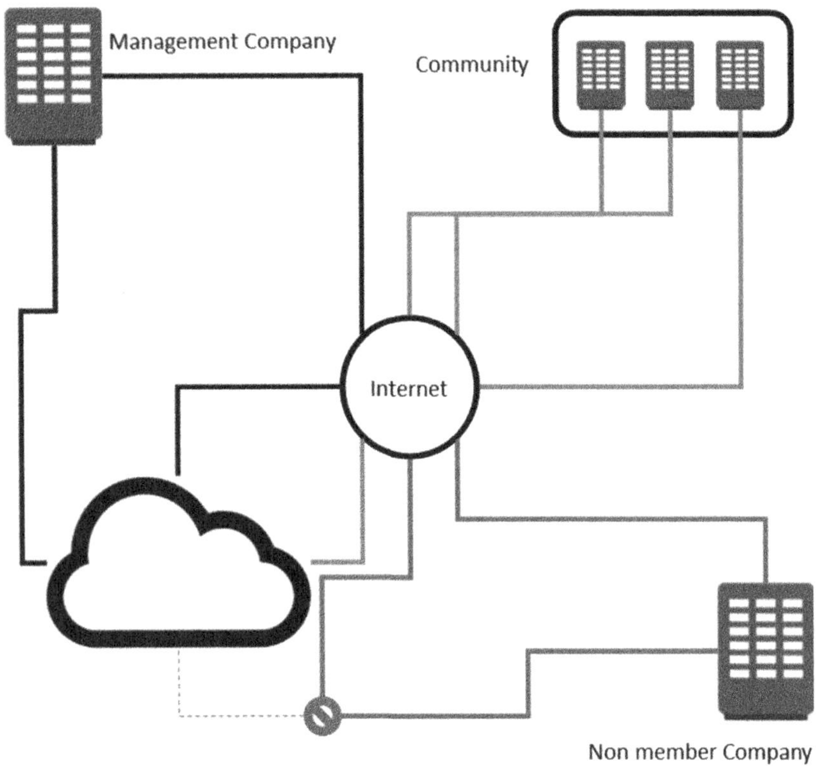

Figure 5.3. *Community deployment of cloud computing. For a color version of this figure, see www.iste.co.uk/benmammar/network.zip*

Figure 5.3 illustrates a community deployment model. The infrastructure can be managed by the company outside the community (managing company), by the entire community or a part of it. Therefore, a group of companies must share a minimum of requirements and/or to draw a maximum of benefits from the model.

5.2.1.2.4. Hybrid model

The infrastructure is a mixture of at least two previously mentioned models (Ahmad *et al.* 2017).

5.2.1.3. *Levels of Cloud Computing services*

5.2.1.3.1. Infrastructure-as-a-service

Accessible and shared resources are provided to users, such as computing units, storage media, and communication means (network) (Mell and Grance 2011). Using these resources, the consumers can remotely run their own software and tools (Hayes 2008). This means that the customer controls the operating system, the applicative aspect and the storage (Syntec Numérique 2012), but does not manage the low layers, such as the infrastructure and the material network (Hayes 2008). Nevertheless, he/she may sometimes partially manage the network software (if authorized by the provider), such as the choice of firewall (Armbrust *et al.* 2010).

5.2.1.3.2. Software-as-a-service

The objective is to use web applications that run on cloud servers (Hayes 2008). The software can be accessed by various customers, across the globe (Mell and Grance 2011). These customers control only those parameters specifically left open by the service provider (Ahmad *et al.* 2017), and have no control on the management of infrastructure, operating system, network, and the application itself (Syntec Numérique 2012).

5.2.1.3.3. Platform-as-a-service

An intermediate level between the two previously mentioned levels is platform-as-a-service (PaaS) (Ahmad *et al.* 2017). The client can deploy customer-created applications on the tools offered or proposed by the provider (Mell and Grance 2011). These may include databases and programming languages (Hayes 2008). The consumer does not manage the lower layers such as servers and operating systems (Syntec Numérique 2012), but is however in control of the deployed applications and is able to configure their hosting environment (Armbrust *et al.* 2010).

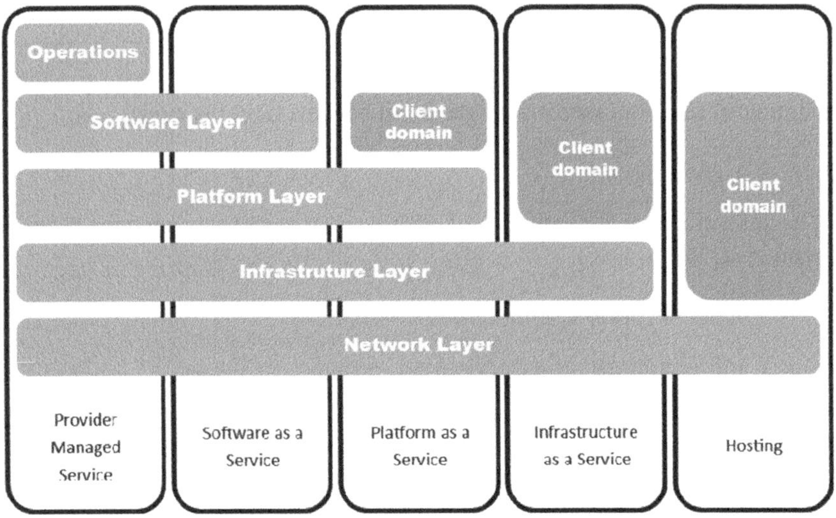

Figure 5.4. *Distribution of customer/provider management in a cloud computing environment depending on service level (Syntec Numérique 2012)*

Figure 5.4 shows the way in which Syntec Numérique (2012) distinguishes between levels of cloud computing services. For example, Facebook and most Google products are the most famous software-as-a-service (SaaS).

PaaS level is quite difficult to understand, given the ambiguity of customer and provider shared control. For example, the customer may control only logic data such as Heroku service, or a real DataBase Management System (DBMS) in a cloud environment such as My Structured Query Language (MySQL) cloud. Finally, the customer may control a whole integrated development environment such as Salesforce (the service gives its customers the possibility to create a complex website in several steps).

Infrastructure-as-a-service (IaaS) level enables the customer to control a remote virtual machine. Consequently, the customer has a distant computer that he/she does not physically possess, but which is always available for him/her.

5.2.2. *Artificial intelligence*

AI refers to intelligence manifest in machines, unlike natural intelligence possessed by human beings or animals (Russel and Norvig 2016).

AI is a science that studies intelligent agents (Legg 2007), which are entities simultaneously capable of perceiving the environment in which they exist and of making decisions in order to maximize the chances of reaching their objectives (Kaplan and Haenlein 2019).

This is the most simple and common definition in the field, as many researchers estimate that people amalgamate everything under the umbrella of AI (McCorduck 2009). Furthermore, this phenomenon is referred to as "AI effect" (McCorduck 2009). As an example, many researchers in the field of AI do not consider optical character recognition as being part of AI. This is due to the deterministic and optimal algorithms solving this type of problems (Schank 1991).

5.2.2.1. *Subjects*

The subjects dealt with in the field of AI are diverse and various. The most widely known is (reasoning-based) problem resolution (Poole *et al.* 1998). Since reasoning is one of the facets of natural intelligence, AI researchers took a step-by-step approach to the implementation of reasoning in machines (Nilsson 1998).

The concept has steadily evolved, but the run time of currently proposed algorithms increases (exponentially) as soon as the problem exceeds a certain level of complexity (Russel and Norvig 2016). It is easier to solve such problems of significant complexity with natural intelligence, because human beings use shortcuts (such as intuition and memory) most of the time (Foss and Dodwell 1966).

Since AI studies intelligent agents able to perceive their environments, it is obvious that perception itself is a theme in the field (Nilsson 1998; Russel and Norvig 2016). Further examples are voice recognition (Russel and Norvig 2016), facial recognition and object recognition (Russel and Norvig 2016), which are subthemes related to the integration of sensors in intelligent agents.

AI covers many themes. Those explained in this section are in no way a representation of the whole range of themes.

Indeed, some problems are better considered as methods (such as machine learning) or applications (such as language recognition).

5.2.2.2. *Approaches*

In order to cope with the previously mentioned problems, AI has a broad range of approaches at its disposal. While several researchers thought that machines should feel emotions, others preferred them to find the very essence of solving logic and varied problems (Crevier 1993; McCorduck 2009) (as knowledge representation, planning and learning).

Further approaches can be mentioned, such as intelligent computing. This studies the integration of artificial neural networks in computing (Crevier 1993; McCorduck 2009) and is used to find low-certainty solutions to problems that do not necessarily require an optimal (global) solution.

AI uses many approaches that are beyond the scope of a chapter that does not deal specifically with the field. Therefore, the approaches that are briefly explained in this section do not represent all the possible approaches in AI. Moreover, certain approaches are better considered as being domains (such as cybernetics) or even subdomains (systems of intelligent agents).

5.2.2.3. *Tools*

Each domain has its problems, and each problem needs tools to be solved.

Among the most commonly used tools in AI, artificial neural networks should be being mentioned. They are heavily inspired by the architecture and behavior of human brain. They are known as a set of interconnected nodes. Each node is associated with a weight in order to vote for a decision it is expected to activate (Domingos 2015). The only drawback of an artificial neural network is that it needs a certain time in order to learn from examples. Furthermore, though less probable, it is not impossible to learn a false decision.

On the other hand, some researchers use probabilistic methods for uncertain reasoning. These are algorithms that do not have all their inputs. These methods include Bayesian networks (Nilsson 1998), Markov models (Russel and Norvig 2016) and Kalman filters (Russel and Norvig 2016).

A description of all the methods used in the selection of cloud computing services is beyond the scope of this section. The reader should find it more interesting to inspect a review (short description) of the currently used methods and contributions to the selection of cloud computing services. Consequently, further details on the tools used by these contributions are provided in section 5.4.

5.3. Similar works

This section offers a brief presentation of already published surveys for cloud computing services selection contributions. After introducing each work, we review it and explain why the paper is recommended.

Beloglazov *et al.* (2012) monitored the heuristics used for the discovery and allocation of datacenter resources in cloud computing environment, while reducing the energy consumption and providing the negotiated quality of service. To achieve this, researchers divided their works along three distinct axes. The first one focuses on computerized cloud architectures enabling energy consumption reduction or control. The second axis is dedicated to planning policies and algorithms that provide a previously determined quality of service, while reducing energy consumption. The third and last axis is more interested in problems than in solutions. It presents the currently open problems while trying to allocate the resources of a computerized cloud, reduce energy consumption and provide the required (negotiated) quality of service.

The above quoted work is strongly recommended for any researcher (either non-initiated, beginner or expert) willing to contribute to resource allocation in computerized clouds.

Some works applied multiple choice decision making methods for IaaS cloud service selection. At first, Rehman *et al.* (2012) deployed a specific number of methods on the same set of IaaS cloud services. Then they compared the results found. In another research paper, the same authors added quality of services to their research criteria (Rehman *et al.* 2014).

The two excellent reports quoted above are of interest for any researcher willing to contribute to cloud computing service selection using neural networks. This is all the more so, as most approaches requiring learning generally rely on multiple-choice decision-making methods.

During the same year, Mandal *et al.* (2013) compared the approaches in order to discover, evaluate and select cloud computing services for large data flow applications. The authors first presented an architecture, and then compared the results of their work with the topical results. It is worth mentioning that all their contributions rely on trees.

Although quite outdated, the respective article is an excellent introduction to the discovery of cloud computing services using intelligent agents.

Le *et al.* (2014) proposed a multicriteria decision-making approach based on fuzzy logic for the selection of the best cloud computing services. For this purpose, researchers proposed an approach based on both interpretive structure modeling (ISM) and analytic network process (ANP). The first served for modeling interactive relations between evaluation criteria, while the second served for the management of uncertain data. Given that the objective was to maximize the useful results, an optimal result was perceived as a service maximizing the utility function. It is worth being mentioned that the authors ignored the time required to get to such results.

Some AI applications focus on the prediction of uncertain data. An uninitiated researcher looking for a contribution from this perspective will find this report highly recommended.

Sun *et al.* (2014) summarized the scientific advances accomplished in 2014 in cloud computing services selection. They then proposed several directions for the near future. The authors ignored the experimental side of the studied works, but they nevertheless summarized the evolution of problems during the period 2009–2014.

In Sun *et al.* 2014, we found an excellent introduction to cloud computing services selection if the researcher does not yet have a clearly defined problem.

Whaiduzzaman *et al.* (2014) proposed a survey of the works selecting cloud computing services by multiple-choice multi-criteria methods (MCMC). To reach their objective, they first drew a synthesis of MCMC techniques (including AI). Then they analyzed the approaches in their study domain. Finally, they highlighted certain aspects of MCMC methods in the selection of cloud computing services. Many of the proven facts rely on AI.

The above-mentioned paper focuses on MCMC, but this type of approach often heavily relies on AI. Consequently, the paper written by Whaiduzzaman *et al.* (2014) is recommended to researchers that are not initiated in the field of AI and are willing to contribute to the selection of cloud computing services.

Papathansiou *et al.* (2015) resumed the work of Whaiduzzaman *et al.* (2014) in order to add an analytical comparison in terms of objectives and performances. Researchers had to implement the surveyed processes to achieve that, and then submit them to evaluation and results comparison. In addition to the article by Whaiduzzaman (2014), Papathanasiou *et al.* (2015) is a stronger recommended reference.

Lastly, Ahmed *et al.* (2019) proposed a survey as well as an analysis of the requirements for the approaches selecting cloud computing services for cloud federation (or federation of cloud computing services). First, researchers started by defining the characteristics of federated computerized clouds, as well as services in federated clouds. Then, they proposed their own definition of level of trust for a cloud computing service. Therefore, they compared their measures of trust to those proposed in the literature. Finally, they highlighted the advantages of their proposal compared to the proposals in the field of the cloud computing services federation.

Although the work of Ahmed *et al.* (2019) looks more like a contribution than a survey, it was considered like a state of the art because the researchers explicitly mention it. However, a researcher may not necessarily be interested in the federation of computerized clouds; this work is recommended for the composition of cloud computing services (general case).

Finally, there is implicitness in the field. Some works are positioned on the customer side, and others on the provider side. The latter deal with billing diagrams (Aishwarya and Muzammil 2017), energy consumption efficiency (Sharma *et al.* 2016b), task planning (Singh *et al.* 2017), profit maximization (Das *et al.* 2014) and resource management (Liaqat 2017) (this list of objectives is not exhaustive).

On the other hand, the works on the customer side rather focus on cost optimization (Pandey *et al.* 2011), migration of applications toward cloud computing (Andrikopoulos *et al.* 2013), risk management (Gupta *et al.* 2015) and data confidentiality (Xu *et al.* 2016) (this list of objectives is not exhaustive).

5.4. Surveyed works

This section covers a certain number of works in the selection of cloud computing services using AI (problems, tools, approaches, fields application, etc.).

To improve its readability, the set of surveyed works is divided into families, according to the used processes (subdomains of AI). The objective is to reduce the number of contributions per subsection, and to facilitate a more detailed discussion of subdomain specific problems.

5.4.1. *Machine learning*

This section reviews work in cloud computing services selection using machine learning for this purpose. It is a known fact that AI is an implementation of human intelligence by machines, and learning by examples is a form of natural intelligence.

Machine learning enables intelligent agents to learn from examples before confronting real cases.

Kang *et al.* (2013) proposed an approach for the selection of a virtual machine in a cloud environment in order to plan and distribute user processes tasks. The objective was to maximize the best virtual machine prediction, as well as its performances, before seeing the price or quality, while learning on a set of examples. Despite the interest of the contribution, learning requires a certain time in order to get to a useful stage; the authors in fact neglected this aspect (time required for learning).

Xiaogang *et al.* (2015) proposed a dynamic approach for the selection of the best cloud computing service. For this purpose, they implemented a strategy known as dynamic cloud service (DCS) in each broker. The strategy relies on machine learning and its objective is real time provision of the best service, irrespective of previous changes. Researchers tested the performance of their contribution in terms of success percentage and failure probability. Although interesting, well explained and presented, the work should be more appropriately evaluated in terms of failure (or success) percentage and time required for taking into account the change in the price of a cloud computing service that provides the best service (since a real-time response is just a very rapid answer).

Sharma *et al.* (2016b) proposed an approach to predict the parameters that support (attract or motivate) cloud computing adoption. As a first step to reach their objective, the researchers improved two techniques, which rely on *multiple linear regressions* and on neural networks. Even though the authors tested the two improved techniques via study cases, they did not find the same order of importance of the parameters (for the same set of criteria at the beginning). Moreover, the authors did not analyze the reasons behind such a difference.

Li *et al.* (2017) proposed a decision method in conjunction with the multiple choice method in order to compose cloud computing services for previously determined purposes (cloud manufacturing). To do so, researchers first determined their selection criteria. In other terms, they stated how a single service would be evaluated. Then, the authors used neural networks, fuzzy logic, hierarchic analysis process, TOPSIS, etc., in order to select, compose and evaluate cloud computing services composition. Although their work is well explained, illustrated and presented, researchers nevertheless ignored a performance evaluation of the proposed contribution, as well as the comparison with other similar approaches. Furthermore, this article is not recommended for a researcher who is not an expert in the used methods, since certain implicitness was detected in relation to combinations of the latter.

Alipoufard *et al.* (2017) proposed a full system selecting IaaS cloud computing instances for big data analyses. To do so, researchers used statistical learning and Bayesian optimization. The work is well explained, nicely illustrated and rigorously presented. Moreover, the authors could experiment on real examples of IaaS cloud computing services and could also evaluate the performances of their approach, but they did not deem it necessary to compare their contribution to similar works. However, the article is still recommended to any researcher interested in cloud computing services selection using statistical learning or Bayesian algorithms.

Wassim *et al.* (2018) proposed an analysis of the factors using unsupervised machine learning in order to evaluate cloud computing services. For this purpose, researchers considered two types of quality metrics, the first being customer feedback and the second server feedback. Then they analyzed the criteria of cloud computing services based on machine learning. They used customer feedback in order to have objective and subjective quality measures. At the end, they were able to rigorously calculate a score for cloud computing services. This contribution is very well explained and the authors' efforts to evaluate their approach are worth mentioning. However, the researchers did not compare their work to similar works.

The works quoted in this section are not an exhaustive list of scientific articles in the selection of cloud computing services using machine learning. Given the wide extent of AI in the world of computer science, it is impossible to mention all the articles, but it can be noted that they all ignore the time required for learning. Researchers are often not interested in evaluating the time required for learning in their approaches. In other terms, researchers have a tendency to evaluate their proposed approaches in terms of success rate (very often), accuracy (quite often), failure rate (not frequent), but they consistently ignore the evaluation of the time required for their approaches to learn.

5.4.2. *Heuristics*

In order to solve problems related to optimization, research or both, certain researchers use heuristics and metaheuristics. It is worth noting that in everyday life human beings eliminate certain cases during their researches in order to save time. Knowing that the eliminated solutions may be correct (valid) and/or optimal, such situations are generally eliminated for lack of time. It is the very principle of heuristics.

Gao *et al.* (2012) proposed a technique for balancing the loads of processes between mobile customers and their services in a computerized cloud. For this purpose, researchers first reviewed several required theoretical concepts (such as Cloudlet, Coordinated Mobile Device, etc.). Then, they presented their model. This has the characteristics of processes, networks, etc. After that, they proposed an algorithm with heuristics aimed at balancing the loads of processes for customers and services. The algorithm produces offloading plans verifying certain previously established utility functions. Heuristics distinguishes between an optimal and a less optimal solution during the research. Taking a look at the contribution performance tests conducted by the authors themselves, it can be noted that the approach was not compared to other similar approaches. Moreover, researchers ignored energy consumption on both sides (customer and server), cost of service in computerized cloud, etc. The work is nevertheless interesting. Consequently, it is recommended to any researcher interested in load balancing (in cloud computing, grid computing, etc.).

Nacsimento *et al.* (2016) proposed an improvement and a facilitation of the scaling of virtual instances in computerized clouds. For this purpose, they first investigated the efficiency of machine learning techniques for scaling up. Then, they proposed a set of heuristics to improve the learning algorithms. However, the objective to be optimized is not very clear (such as the time for starting the virtual instance, maximization of the number of virtual machines for each physical machine, etc.). It is also worth noting that the work was neither compared, nor evaluated, but the authors made an effort to thoroughly explain the problem and conduct a case study.

Hoang *et al.* (2016) proposed an approach to customer query management in a cloud computing environment, relying on ant colony heuristics and the *particle swarm optimization* (PSO). For this purpose, researchers first established several objectives (they can play the role of utility functions) as follows: system overall cost minimization, meeting the measures of quality of service and profit maximization for cloud computing services providers. The work is well explained, nicely illustrated, evaluated, but it was not compared to other similar approaches. The article is nevertheless interesting and it is highly recommended to researchers interested in customer query planning, scheduling and management in a cloud computing environment.

Similarly, Xue *et al.* (2016) proposed an approach to process scheduling in a computerized cloud. Relying on PSO and heuristics, researchers set a single objective, namely maximization of the quality of service. The authors made an effort to explain their work, evaluate their approach and compare their contribution to similar ones. However, researchers expressed their objective (maximization of the quality of service) in terms of minimization of the total time of customer query execution, while the quality of service can be the minimization of the average time for customer task completion.

Moreover, Samieifar and Mardukhi (2017) proposed an approach aiming at the dynamic resource allocation in a computerized cloud. Assuming the problem was NP-Hard, researchers proposed a metaheuristic combining genetic algorithms and colonial competition. The authors explained the concepts very well, appropriately illustrated the architecture and the research context, and compared their contribution to similar works. Even though the authors took into account the execution time and the cost of services, a researcher can evaluate its approach in terms of services resources consumption. Consequently, this work is recommended to any researcher willing to contribute to the allocation of resources in computerized clouds.

Finally, Hajlaoui *et al.* (2017) proposed a system discovering and selecting IaaS cloud computing services. For this purpose, they introduced two heuristics, namely Hungarian and Volgenant-Jonker algorithms. Then, they used the graph theory, linear programming, as well as symmetric linear transformation in order to calculate the quality of the discovered services. Finally, to prove the efficiency of their approach, the authors compared the employed concepts, instead of the proposed approaches. The article is however interesting, and recommended for any researcher willing to contribute to the discovery and/or selection of cloud computing services.

This section presented several works using heuristics or metaheuristics for the selection, evaluation, composition or discovery of cloud computing services. This list of contributions is far from being exhaustive, but offers examples of ideas for future researches. Moreover, it is worth noting that works proposing one or more heuristics ignore one aspect. In general, researchers do not compare the approaches using their proposed heuristics to approaches using standard heuristics (already proposed in the literature). The purpose is to separately evaluate the heuristics and the algorithm.

5.4.3. *Intelligent multiagent systems*

Intelligent multiagent systems are the first alternative used by AI researchers in studying the impact one agent has on the others. As an example, let us consider N intelligent customer agents and M intelligent agents providing cloud computing services. Let us assume that among the proposed services, a provider has a prominent position due to its optimal quality and reduced price (denoted by m_i). The demands of N customer agents for the service offered by m_i are increasing.

Consequently, the provider needs to take precautions (or measures, in general) to prevent their servers from being overloaded. For example, it can add material servers (which tends to increase the price), or increase the number of virtual servers for each physical machine (which tends to lower the quality of service), etc. In all cases, the measures taken may affect the customers, who may choose another provider.

This is a simple example among others of intelligent multiagent systems applications in the selection of cloud computing services. This section presents several contributions using AI and multiagent systems for the discovery, selection, evaluation or composition of cloud computing services.

Rabbani *et al.* (2014) proposed a method that uses intelligent agents for the selection of cloud computing services. As researchers noted, if sufficiently trained, the agents would at the end be able to observe the similarities and differences between services and select them. Given that the objective was to best meet the customer requirements, it would be interesting to see the evaluation of such an approach in terms of learning time, relevance of results, etc., but the authors do not provide an evaluation or comparison of their work.

Chichin *et al.* (2014) proposed an intelligent agent-based market for cloud computing services. Their principle relies on the simulation of cases by intelligent agents (software) according to several policies, various situations, etc. If a customer wants to use the proposed platform, the authors recommend the agent that is best adapted to its needs and situation.

This work is quite remarkable, given that researchers evaluated their approach and compared it to several other similar contributions. Moreover, the comparison criteria are diverse and varied. This article is highly recommended to any researcher willing to contribute to the selection of cloud computing services using intelligent agents.

Lacheheub *et al.* (2016) proposed an approach to cloud computing services selection, relying on intelligent agents. First, the agents discover cloud computing services gradually. Then, each agent hierarchically groups them. Therefore, the customers select services that are closest to their former services (provider change) or their local applications (migration to computerized clouds).

Among all surveyed contributions, Lacheheub and Maamrin (2016) is the best in terms of explanation, illustrations and evaluation. Besides explaining and illustrating their approach, the authors studied an example, unfolded the process to be followed (from mathematical, graphical and linguistic perspectives), evaluated their approach and compared it to other contributions. Consequently, this article is recommended to any researcher willing to contribute to the selection of cloud computing services using intelligent agents.

Jahani *et al.* (2017) proposed a system known as ARank, selecting the cloud computing services and relying on intelligent multiagents system. First, each agent takes several cloud computing services. Then, each agent evaluates its candidates according to the quality of service. Their evaluation includes the customer level of satisfaction history for each cloud computing service. The principle is easy to understand, as the authors thoroughly explain the concept. However, there is a certain inconsistency between objective and evaluation (comparison between the proposed approaches, Analytic Hierarchy Process (AHP)) Singular Value Decomposition (SVD). As the researchers state, the customer's wait time is optimized, as they compare their contribution to others in terms of evaluation time with respect to the total number of cloud computing services. In this case, the reader may understand that services are not evaluated at the selection moment, but that the opposite is true.

5.4.4. *Game theory*

Mathematical and/or computing problems are often expressed in the form of games following a mathematical model for the study of decision makers in their environments. The predator–prey model is a good start in understanding the interest of game theory. Let us consider the example of a determined area containing a predator set and a prey set. It can be immediately noted that the two groups of players are competing. For example, lions hunt buffalos and they are competing. Lions cannot do without hunting, as it is a matter of survival. Buffaloes have to protect themselves to avoid extinction or herd weakness. Therefore, the predator is always confronted with its prey resistance, to which they are beneficial. If buffaloes stopped defending themselves, the lions would more easily and frequently attack the herd. Consequently, the herd would be weakened or become extinct. In these cases, if the buffalo do not become extinct, they will be forced to migrate to a safer place. This renders them vulnerable during the trip and weakens the predating lions (the latter have fewer sources of nutrition). On the other hand, if the lions stopped hunting buffaloes, they would certainly become extinct and buffaloes would have more chances to reproduce and strengthen their herd. But, buffaloes are also predators for other species in the studied area. If their sources of nutrition become insufficient, they would also become extinct.

Game theory is not limited to the simple study of the decision maker (either in a game or in reality). It also studies the impact of a decision on the players (the other decision makers) in the same environment (case of sharing material resources in a computerized cloud).

Hassan *et al.* (2014) proposed an approach allocating cloud computing resources in a context of dynamic horizontal federation of computerized clouds. For this purpose, the researchers first presented the architecture of a system for horizontal and dynamic federation of cloud computing services. Then they studied two cases of utility function maximization for resource allocation according to the presented context. They next proposed an algorithm that selects resources depending on their prices. The work is well presented, illustrated and explained, but the authors did not find it necessary to include a comparison of their approach with already existing contributions. Moreover, their evaluation (utility function maximization, social well-being maximization, minimization of the number of virtual machines per cloud computing provider, etc.) does not match the objectives of their approach (robustness maximization and time reduction).

Do *et al.* (2015) presented an interclient cooperation model for the selection of the best cloud computing service. For this purpose, they first divided the set of customers into groups. Then, they set things such that each customer randomly chooses a provider at the start. After that, each customer communicates the characteristics of his/her provider to other customers that belong to the same group. At this moment, the discovered phase is completed and the customers have to change their provider if they are communicated the characteristics of a better one. Given that the study aimed to present a comprehensive definition of the heterogeneous cloud computing services markets, in our opinion the objective was largely fulfilled, as the researchers proposed a mathematical modeling of the problem, as well as a numerical evaluation. Therefore, the authors did not deem it necessary to compare their approach to other similar works (if they still exist).

Ardagna *et al.* (2015) proposed an approach to income maximization and cost minimization for providers of SaaS cloud computing services, who in turn are customers of IaaS cloud computing services. For this purpose, they proposed a *Nash* game model in which providers of SaaS cloud computing services bet on the resources allocated at IaaS level. Researchers could prove the existence of a certain *Nash equilibrium point*, and the fact that convergence occurs in a limited number of iterations. Therefore, the authors mentioned that they compared their approach to other similar techniques in the state of the art, but we could not find it in the paper. The researchers just evaluated the quality of service during a certain lapse of time.

Liu *et al.* (2016) proposed a study of the impact of invoicing model communication on the speed of convergence toward equilibrium state in a provider/customer environment in a computerized cloud. For this purpose, researchers first formalized the problem and presented the Stackelberg game model. Then, they presented strategies that are optimal for both customers and providers in order to maximize their utility (the authors also proved such strategies are optimal). Therefore, they could reach the Stackelberg equilibrium point by unfolding optimal strategies presented in a computerized cloud environment. Finally, the researchers modified several parameters of the simulation in order to see the impact on the equilibrium and the time required for its attainment. Nevertheless, a comparison with similar approaches or at least a Nash game-based approach would be highly appreciated.

Wu *et al.* (2016) proposed a mechanism for resource allocation in self-organized computerized clouds. For this purpose, the researchers proposed two new economic strategies allocating resources based on architecture and prices. More specifically, they used the modified Vickrey double auction when resources were sufficient and the continuous double auction when resources were insufficient. Despite the authors' remarkable explanation, they did not deem it necessary to compare their approach with other similar techniques. Furthermore, the researchers evaluated their contribution in terms of cost for resource procurement with respect to the number of providers and execution efficiency. However, there are many other important parameters such as procurement delay, the number of virtual machines compared to the number of physical machines, etc.

This section presented a set of works in the selection of cloud computing services relying on AI in game theory. It is worth noting that collective intelligence is involved, since the majority of contributions consider customers and/or providers as players. Each has an objective to achieve and they must share the same material resources. In technical terms, researchers tried first to prove the existence of a certain point of equilibrium (explicitly or implicitly). Then they tried to find a way to reach it. Finally, the authors simply had to prove that the equilibrium point could be reached through a finite number of iterations (a more or less limited time). Some untapped research opportunities were highlighted, such as an equilibrium point not being manifest even though it previously existed, the heterogeneity of players (customers and/or providers), their final objectives, their strategies, etc.

Genetic algorithms are another example of AI techniques that were used for the selection of cloud computing services. They include algorithms such as the ant colonies algorithm (Gao 2014), the bee colonies algorithm (Tian *et al.* 2013; Seghir *et al.* 2016; Xu *et al.* 2017; Zhou and Yao 2017), the crow search algorithm (Satpathy *et al.* 2017), etc.

AI covers a broad range of methods, techniques and processes, in order to best address its problems. It is beyond the scope of this chapter to mention all the contributions in the field, or even one contribution per each subdomain of AI.

5.5. Conclusion

This chapter presents a state of the art in the selection, composition, evaluation or recommendation of cloud computing services using AI. It offers an overview of the problems considered, approaches followed and tools used in the two domains, with the mention that such survey studies are seldom precise and/or detailed.

Our near-future objective is to focus on a comparison in addition to the survey so that the reader would have access to the performance test results of the surveyed approaches (besides the critical review).

A further objective is to integrate the limits of AI in the selection of cloud computing services. In other words, it is our intent to integrate the impact of AI limits on the surveyed studies.

Moreover, given the broad range of game theory-based works, a more in-depth survey of the latter is intended. As an example, the bidirectional bet technique is often used in game theory through AI.

5.6. References

Ahmad, I., Bakht, H., and Mohan, U. (2017). Cloud computing – A comprehensive definition. *Journal of Computing and Management Studies (JCMS)*. 1(1), 1–8.

Ahmed, U., Raza, I., and Hussain, S.A. (2019). Trust evaluation in cross-cloud federation: Survey and requirement analysis. *ACM Computing Surveys (CSUR)*, 52(1), 19.

Aishwarya, S. and Muzammil, H. (2017). Pricing schemes in cloud computing: a review. *International Journal of Advanced Computer Research*, 7(29), 60.

Alipourfard, O., Liu, H. H., Chen, J., Venkataraman, S., Yu, M., and Zhang, M. (2017). Cherrypick: Adaptively unearthing the best cloud configurations for big data analytics. *Symposium on Networked Systems Design and Implementation (NSDI)*. 27–29 March. Boston. USA.

Andrikopoulos, V., Song, Z., and Leymann, F. (2013). Supporting the migration of applications to the cloud through a decision support system. *IEEE Sixth International Conference on Cloud Computing*. IEEE, Santa Clara, 565–572.

Ardagna, D., Ciavotta, M., and Passacantando, M. (2015). Generalized Nash equilibria for the service provisioning problem in multi-cloud systems. *IEEE Transactions on Services Computing*, 10(3), 381–395.

Beloglazov, A., Abawajy, J., and Buyya, R. (2012). Energy-aware resource allocation heuristics for efficient management of data centers for cloud computing. *Future Generation Computer Systems*, 28(5), 755–768.

Chichin, S., Chhetri, M.B., Vo, Q.B., Kowalczyk, R., and Stepniak, M. (2014). Smart cloud marketplace-agent-based platform for trading cloud services. *IEEE/WIC/ACM International Joint Conferences on Web Intelligence (WI) and Intelligent Agent Technologies (IAT)*. IEEE/WIC/ACM, Warsaw, Poland.

Crevier, D. (1993). *AI: The Tumultuous History of the Search for Artificial Intelligence*. Basic Books, New York.

Das, A.K., Adhikary, T., Razzaque, M.A., Cho, E.J., and Hong, C.S. (2014). A QoS and profit aware cloud confederation model for IaaS service providers. *8th International Conference on Ubiquitous Information Management and Communication*. ACM, Siem Reap.

Do, C.T., Tran, N.H., Huh, E.N., Hong, C.S., Niyato, D., and Han, Z. (2015). Dynamics of service selection and provider pricing game in heterogeneous cloud market. *Journal of Network and Computer Applications*, 69, 152–165.

Domingos, P. (ed.) (2015). How does your Brain Learn? *The Master Algorithm: How the quest for the ultimate learning machine will remake our world*. Basic Books, New York.

Foss, B. and Dodwell, P.C. (1966). *New Horizons in Psychology*. Penguin Books, London.

Gao, B., He, L., Liu, L., Li, K., and Jarvis, S.A. (2012). From mobiles to clouds: Developing energy-aware offloading strategies for workflows. *ACM/IEEE 13th International Conference on Grid Computing*. ACM/IEEE, Beijing.

Gao, Z. (2014). The allocation of cloud computing resources based on the improved ant colony algorithm. *Sixth International Conference on Intelligent Human-Machine Systems and Cybernetics*. IHMSC, Hangzhou.

Gupta, S., Muntes-Mulero, V., Matthews, P., Dominiak, J., Omerovic, A., Aranda, J., and Seycek, S. (2015). Risk-driven framework for decision support in cloud service selection. *15th IEEE/ACM International Symposium on Cluster, Cloud and Grid Computing*. IEEE/ACM, Shenzhen.

Hajlaoui, J.E., Omri, M.N., Benslimane, D., and Barhamgi, M. (2017). QoS based framework for configurable Iaas cloud services discovery. *IEEE International Conference on Web Services (ICWS)*. IEEE, Hawaii.

Hassan, M.M., Hossain, M.S., Sarkar, A.J., and Huh, E.N. (2014). Cooperative game-based distributed resource allocation in horizontal dynamic cloud federation platform. *Information Systems Frontiers*, 16(4), 523–542.

Hayes, B. (2008). Cloud computing. *Communications of the ACM*, 51, 9–11.

Hoang, H.N., Le Van, S., Maue, H.N., and Bien, C.P.N. (2016). Admission control and scheduling algorithms based on ACO and PSO heuristic for optimizing cost in cloud computing. In *Studies in Computational Intelligence*, Hoang, H.N., Le, V.S. *et al.* (eds). Springer, Berlin.

Jahani, A., Derakhshan, F., and Khanli, L.M. (2017). A Rank: A multiagent based approach for ranking of cloud computing service. *Scalable Computing: Practice and Experience*, 18(2), 105–116.

Kang, D.S., Liu, H., Singh, M.P., and Sun, T. (2013). Adaptive process execution in a service cloud: Service selection and scheduling based on machine learning. *IEEE 20th International Conference on Web Services*. IEEE, Santa Clara.

Kaplan, A. and Haenlein, M. (2019). Siri, Siri, in my hand: Who's the fairest in the land? On the interpretations, illustrations, and implications of artificial intelligence. *Business Horizons*, 62(1), 15–25.

Lacheheub, M.N. and Maamrin, R. (2016). Towards a construction of an intelligent business process based on cloud services and driven by degree of similarity and QoS. *Information Systems Frontiers*, 18(6), 1085–1102.

Le, S., Dong, H., Hussain, F.K., Hussain, O.K., Ma, J., and Zhang, Y. (2014). Multicriteria decision making with fuzziness and criteria interdependence in cloud service selection. *International Conference on Fuzzy Systems (FUZZ-IEEE)*. IEEE, Beijing.

Legg, S., and Hutter, M. (2007). A collection of definitions of intelligence. *Frontiers in Artificial Intelligence and Applications*, 157, 17.

Li, L., Hang, J., Sun, H., and Wang, L. (2017). A conjunctive multiple-criteria decision-making approach for cloud service supplier selection of manufacturing enterprise. *Advances in Mechanical Engineering*, 9(3), 1–15.

Liaqat, M., Chang, V., Gani, A., Ab Hamid, S.H., Toseef, M., Shoaib, U., and Ali, R.L. (2017). Federated cloud resource management: Review and discussion. *Journal of Network and Computer Applications*, 77, 87–105.

Liu, C., Wang, S., Wang, C., Bie, R., and Shin, D. (2016). Stackelberg game based optimal workload allocation and pricing mechanism in crowdsourcing. *2016 IEEE International Conferences on Big Data and Cloud Computing (BDCloud), Social Computing and Networking (SocialCom), Sustainable Computing and Communications (SustainCom) (BDCloud-SocialCom-SustainCom)*. IEEE, Atlanta.

Mandal, A.K., Changder, S., and Sarkar, A. (2013). Selection of services for data-centric cloud applications: A QoS based approach. *International Conference on Advanced Computing, Networking and Security*. IEEE, Mangalore, 102–107.

McCorduck, P. (2009). *Machines Who Think: A Personal Inquiry into the History and Prospects of Artificial Intelligence*. AK Peters/CRC Press, Natick.

Mell, P. and Grance, T. (2011). The NIST definition of cloud computing. *Computer Security Division, Information Technology Laboratory, National Institute of Standards and Technology*, 800(145), 1–7.

Michael, A., Armando, F., Rean, G., Anthony, J., Randy, K., Andy, K., Gunho, L., David, P., Ariel, R., Ion, S., Matei, Z. (2010). A view of cloud computing. *Communications of the ACM*, 53(4), 50–58.

Nascimento, D.C., Pires, C.E., and Mestre, D.G. (2016). Applying machine learning techniques for scaling out data quality algorithms in cloud computing environments. *Applied Intelligence*, 45(2), 530– 548.

Nilsson, N.J. (1998). *Artificial Intelligence: A New Synthesis*. Morgan Kaufmann, Burlington.

Pandey, G., Rao, V. N., Srivastava, A. K., Banerjee, U., and Narasimhan, E. (2011). Current cloud scenario review and cost optimization by efficient resource provisioning. *The Fourth Annual ACM Bangalore Conference*. ACM, Bangalore.

Papathanasiou, J., Kostoglou, V., and Petkos, D. (2015). A comparative analysis of cloud computing services using multicriteria decision analysis methodologies. *International Journal of Information and Decision Sciences*, 7(1), 51–70.

Poole, D., Mackworth A., Goebel R. (1998). *Computational Intelligence: A Logical Approach*. Oxford University Press, New York.

Rabbani, I.M., Muhammad, A., and AM, M.E. (2014). Intelligent cloud service selection using agents. *9th International Conference on Computing and Information Technology (IC2IT2013)*. IC2IT, Bangkok.

Rehman, Z.U., Hussain, O.K., and Hussain F.K. (2012). Iaas cloud selection using MCDM methods. *IEEE 9th International Conference on e-Business Engineering*. IEEE, Hangzhou.

Rehman, Z.U., Hussain, O.K., and Hussain, F.K. (2014). Time series QoS forecasting for management of cloud services. *Ninth International Conference on Broadband and Wireless Computing, Communication and Applications*. IEEE, Guangdong.

Russel, S.J. and Norvig, P. (2016). *Artificial Intelligence: A Modern Approach*. Pearson Education Limited, Harlow.

Samieifar, S. and Mardukhi, F. (2017). Dynamic resource allocation in cloud computing using a combination of meta-heuristic algorithms. *International Journal of Computer Science and Network Security (IJCSNS)*, 17, 332.

Satpathy, A., Addya, S.K., Turuk, A.K., Majhi, B., and Sahoo, G. (2017). A resource aware VM placement strategy in cloud data centers based on crow search algorithm. *4th International Conference on Advanced Computing and Communication Systems*. ICACCS, Coimbatore.

Schank, R.C. (1991). Where's the AI? *AI Magazine*, 12(4), 38–38.

Seghir, F., Khababa, A., Gaber, J., Chariete, A., and Lorenz, P. (2016). A new discrete imperialist competitive algorithm for QoS-aware service composition in cloud computing. *The International Symposium on Intelligent Systems Technologies and Applications*. ISTA, Jaipur.

Sharma, S.K., Al-Badi, A.H., Govindaluri, S.M., and Al-Kharusi, M.H. (2016a). Predicting motivators of cloud computing adoption: A developing country perspective. *Computers in Human Behavior*, 62, 61–69.

Sharma, Y., Javadi, B., Si, W., and Sun, D. (2016b). Reliability and energy efficiency in cloud computing systems: Survey and taxonomy. *Journal of Network and Computer Applications*, 74, 66–85.

Singh, P., Dutta, M., and Aggarwal, N. (2017). A review of task scheduling based on meta-heuristics approach in cloud computing. *Knowledge and Information Systems*, 52(1), 1–51.

Sun, L., Dong, H., Hussain, F.K., Hussain, O.K., and Chang, E. (2014). Cloud service selection: State-of-the-art and future research directions. *Journal of Network and Computer Applications*, 45, 134–150.

Syntec Numérique (2012). Cloud Computing: nouveaux modèles! White Book, Syntec Numérique, Paris.

Tian, S., Liu, Q., Xu, W., and Yan, J. (2013). A discrete hybrid bees algorithm for service aggregation optimal selection in cloud manufacturing. *14th International Conference on Intelligent Data Engineering and Automated Learning*. IDEAL, Hefei.

Wang, X., Cao, J., and Xiang, Y. (2015). Dynamic cloud service selection using an adaptive learning mechanism in multi-cloud computing. *Journal of Systems and Software*, 100, 195–210.

Wasim, M.U., Ibrahim, A.A., Bouvry, P., and Limba, T. (2018). Cloud service providers optimized ranking algorithm based on machine learning and multi-criteria decision analysis. *Preprints*, 2018010125 (doi: 10.20944/preprints201801.0125.v1).

Whaiduzzaman, M., Gani, A., Anuar, N.B., Shiraz, M., Haque, M.N., and Haque, I.T. (2014). Cloud service selection using multicriteria decision analysis. *The Scientific World Journal* (TSWJ). 459375, 1–10.

Wu, X., Liu, M., Dou, W., Gao, L., and Yu, S. (2016). A scalable and automatic mechanism for resource allocation in selforganizing cloud. *Peer-to-Peer Networking and Applications*, 9(1), 28–41.

Xu, X., Liu, Z., Wang, Z., Sheng, Q. Z., Yu, J., and Wang, X. (2017). S-ABC: A paradigm of service domain-oriented artificial bee colony algorithms for service selection and composition. *Future Generation Computer Systems*, 68, 304–319.

Xu, Y., Li, M., Hu, X., Wang, Y., and Zhang, H. (2016). Survey on privacy preserving for intelligent business recommendation in cloud. In *Wireless Communications, Networking and Applications*, Shi, G. and Ming, Y. (ed.). Springer India, New Delhi.

Xue, S., Shi, W., and Xu, X. (2016). A heuristic scheduling algorithm based on PSO in the cloud computing environment. *International Journal of u- and e-Service, Science and Technology*, 9, 349–362.

Zhou, J. and Yao, X. (2017). A hybrid artificial bee colony algorithm for optimal selection of QoS-based cloud manufacturing service composition. *The International Journal of Advanced Manufacturing Technology*, 88(9/12), 3371–3387.

6

Intelligent Computation Offloading in the Context of Mobile Cloud Computing

Zeinab MOVAHEDI
Iran University of Science and Technology, Tehran, Iran

6.1. Introduction

Nowadays, due to the progress of mobile technologies in terms of both software and hardware, mobile devices, whose use is on the risc, arc becoming an intrinsic part of our everyday life (Li *et al.* 2014). Given these advances, the role of mobile phones has evolved from a simple communication device to an essential tool for many other applications addressing our various daily needs. Examples of such applications include simulations, compression/decompression, image processing, virtual reality, video games, etc. Moreover, in response to the high expectations of mobile users, future mobile devices will use increasingly sophisticated applications. The development of these new applications is, nevertheless, restricted by mobile device limitations in terms of storage space, computation power and battery lifetime (Chen *et al.* 2015).

On the other hand, recent advances in telecommunication networks, in terms of data transmission throughput as well as number of users taken on, offer an opportunity for computation and greedy data offloading to cloud computing and storage center. This latter paradigm, known as mobile cloud computing (MCC), may give rise to a significant number of emerging applications, which are either not possible given the current performance of mobile devices, or are feasible, but require

Intelligent Network Management and Control,
coordinated by Badr BENMAMMAR.

significant processing time and battery energy consumption, therefore frequent recharging of the mobile device (Abolfazli *et al.* 2014; Khan *et al.* 2014).

NOTE.– Mobile cloud computing is a new mobile technology paradigm, in which the capacities of mobile devices are extended using the resources of cloud data centers and computing.

Nevertheless, the efficiency of data and computation offloading to the cloud strongly depends on radio connection quality. The latter is essentially variable in space and time in terms of signal quality, interference, transmission throughput, etc. This dynamic context requires the use of an offloading decision mechanism in charge with determining if a data and computation offloading is beneficial. The offloading benefit is generally evaluated in terms of completed run time and amount of energy consumed, while considering the status of the underlying network (Zhang *et al.* 2016). Moreover, given the overlap of mobile networks such as WIMAX and LTE with wireless local area networks (WAN) such as Wi-Fi and femtocell, as well as the multi-homing property of current mobile devices, the offloading decision mechanism could also contribute to the selection of the appropriate access network for the transmission of the offloaded task data (Magurawalage *et al.* 2015).

Moreover, given that cloud service providers are competing to attract more customers, they may play a role on the MCC market, leading to a multi-cloud environment for mobile device users. In such an environment, the offloading decision mechanism should determine the appropriate cloud provider based on their quality of service and on the cost of computing and storage resources dedicated to the user's query. Market price may however vary depending on supply and demand conditions and interprovider competition (Hong and Kim 2019).

This dynamic context of multi-access networks and multi-clouds requires the enhancement of offloading decision mechanisms by artificial intelligence (AI) tools, enabling a multicriteria decision that best responds to the needs of mobile users in the real world. In this context, this chapter deals with various AI applications for offloading efficiency optimization from the perspective of mobile users, cloud providers and access network providers.

The structure of the rest of this chapter is as follows: Section 6.2 defines the basic offloading notions; section 6.3 presents the MCC architecture in classical access networks and also in cloud-radio access networks; a detailed presentation of the offloading decision is then presented in section 6.4; section 6.5 deals with proposed AI-based solutions to solve the offloading problem; finally, section 6.6 concludes this chapter and proposes several future research directions.

6.2. Basic definitions

As previously described, computation offloading relies on cloud servers that process computations in order to optimize the run time of the mobile application and the energy consumed for task execution.

Given the mobility of the mobile user, the dynamic nature of the radio channel and the variable cost of storage resources and cloud processing, offloading full computations may not always be the best choice. For example, in cases with low bandwidth of the access network, it is more interesting to offload only a subpart of the complete task that corresponds to greedy computation but has very light input data.

NOTE.– Offloading the entire computation may, in some situations, not lead to optimal run time and energy consumption.

Consequently, fine-grain or coarse-grain computation offloading can be done depending on the respective task characteristics and on the surrounding context (Khan *et al.* 2015; Wu 2018).

The two types of offloading are described in further detail in the following sections.

6.2.1. *Fine-grain offloading*

DEFINITION.– *Fine-grain offloading involves the outsourcing of only one subpart of the original task to the cloud.*

In order to determine the subpart of the task whose outsourcing optimizes the application run time and the energy consumed by the mobile device, task components must first be extracted. This can be done based on granularity level, such as class, object and thread. Independently of the chosen granularity level, application components can be extracted with a static code analysis approach or with a dynamic code analysis approach. In the static code analysis, application components are extracted without code execution, while dynamic code analysis relies on the extraction of application components during code execution.

Static code analysis makes it possible to extract components and build the weighted relation graph (WRG) of the application (as described below) prior to the user making an offloading decision. Consequently, the time required for reaching the offloading decision result is shorter. Nevertheless, using static code analysis to extract the application components leads to a less accurate result, particularly for certain types of granularity, for which the program components could vary from one execution to another.

EXAMPLE.– If offloading granularity is considered per object, dynamic code analysis is more appropriate for extracting the application components, as code objects can be created and destroyed depending on the conditions during the application execution.

The WRG of the application can be built from the application components extracted by static or dynamic analysis. It is a graph $WRG = (V, E, W_V, W_E)$, in which each vertex $v \in V$ represents a component of the application and each edge $e \in E$ describes the invocation between the two endpoint components. Each $w_v \in W_V$ and $w_e \in W_E$ represents, respectively, the weight of a vertex v and of an edge e of the graph. Each vertex and edge of this graph is, respectively, weighted according to the number of instructions in the corresponding component and to the amount of data transmitted between the two adjacent components.

An example of application code and the corresponding WRG are illustrated in Figure 6.1.

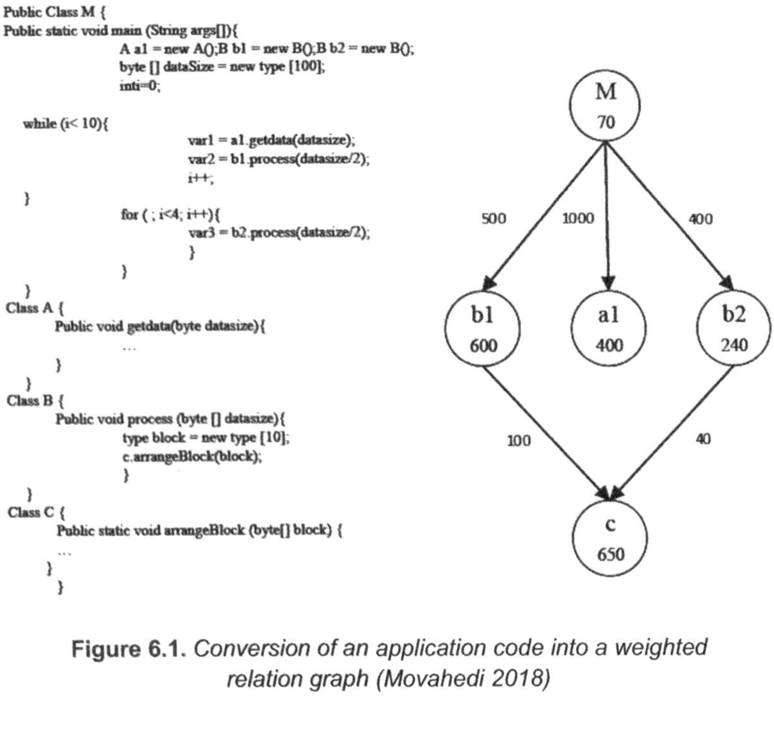

Figure 6.1. *Conversion of an application code into a weighted relation graph (Movahedi 2018)*

Based on the WRG, the fine-grain offloading decision algorithm should determine the components of the application whose offloading ensures optimal run time and energy consumption. This is a multicriteria decision whose efficiency depends on its context awareness, including the underlying radio and cloud environments.

Figure 6.2 illustrates the conceptual model describing the various stages required for making the fine-grain offloading decision.

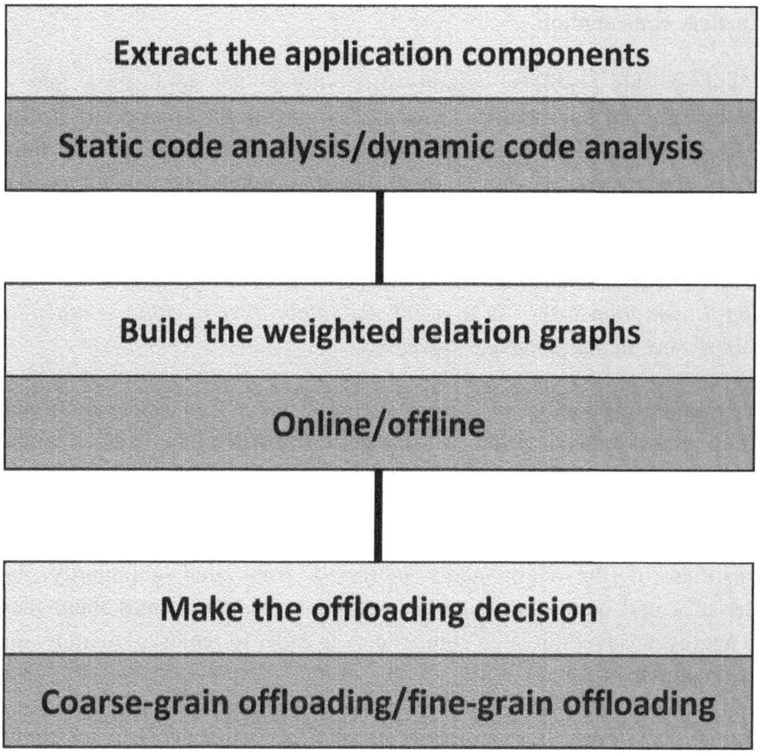

Figure 6.2. *Stages of offloading decision*

6.2.2. *Coarse-grain offloading*

Despite the efficiencies of fine-grain offloading, some works prefer either to offload the entire application or not to offload a program at all. This latter approach, known as "coarse-grain offloading", is motivated by the decision simplicity compared to fine-grain offloading.

DEFINITION.– *Coarse-grain offloading involves outsourcing the execution of an entire program to the cloud.*

NOTE.– Coarse-grain offloading is beneficial especially when a program cannot be divided into several components.

Despite the decision rapidity enabled by coarse-grain offloading, this approach wastes the advantages related to offloading application subparts. These advantages relate to the optimization of the application computing time as well as to the mobile device battery consumption.

To address this problem, certain works divide the application into several subtasks that are not interrelated. This enables making the appropriate decision for each component of the application without having to manage the impact of invocations between components. If these works are taken into account, coarse-grain offloading could be redefined as follows:

DEFINITION.– *Coarse-grain offloading is more generically defined as individual offloading of program tasks. As a result, the application is either considered in its entirety or as several independent components.*

Coarse-grain offloading of several components of an application makes it possible to preserve the simplicity advantage of the offloading decision while at the same time benefiting of an individual decision for each component of the program, and thus improving the resulting efficiency. Nevertheless, dividing the program into several unrelated components is complicated and depends on offloading granularity, functionalities of the components extracted from this granularity and the characteristics and connections of these components. Sometimes there may be a need to aggregate several related components in order to represent them by only one combined component.

A further possibility is to normally extract the components of an application, relying on the chosen granularity, but to represent the connection between two components via the intermediary of its main program, which is obviously implemented in the mobile device. In other terms, each component takes its input parameters from the mobile device and returns its outputs to the mobile device, which then plays the role of intermediary to invoke the connected component.

Processing involves the conversion of the weighted relations graph into a bidirectional star graph. For this purpose, the main component plays the role of the central vertex of the star graph, and the application components play the role of surrounding vertices, each of which is connected to the central vertex by an input edge and an output edge. The input edge with one component designates the input invocation(s) of one or several other components for which the main component acted as intermediary. It is weighted by the sum of weights of the components invoking this component.

Similarly, the output edge of a component represents the output results of this component transmitted to the main in order to be retransmitted to the related component(s). The weight of the output edge is the sum of the weights of the WRG edges receiving an input from this component.

REFRESHER.– A star graph is a connected graph all of the vertices but one of which are degree 1. It can also be seen as a tree with one node and k leaves, at least when $k > 1$.

Figure 6.3 illustrates the weighted star graph corresponding to the relation graph in Figure 6.1.

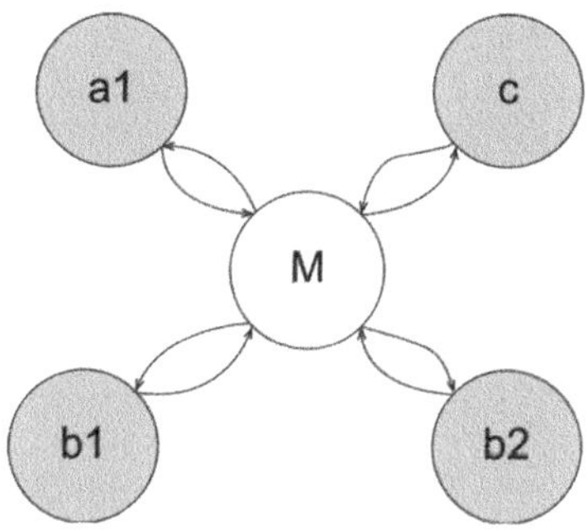

Figure 6.3. *Star graph*

6.3. MCC architecture

In order to clarify the offloading decision ecosystem, this section describes the generic architecture of MCC. Moreover, it presents the C-RAN-based architecture of MCC.

6.3.1. *Generic architecture of MCC*

As illustrated in Figure 6.4, the generic architecture of MCC is composed of five basic elements, namely the mobile device, the access network, the backhaul and backbone network, the cloud and the offloading decision middleware. A detailed presentation of each of these elements is given below.

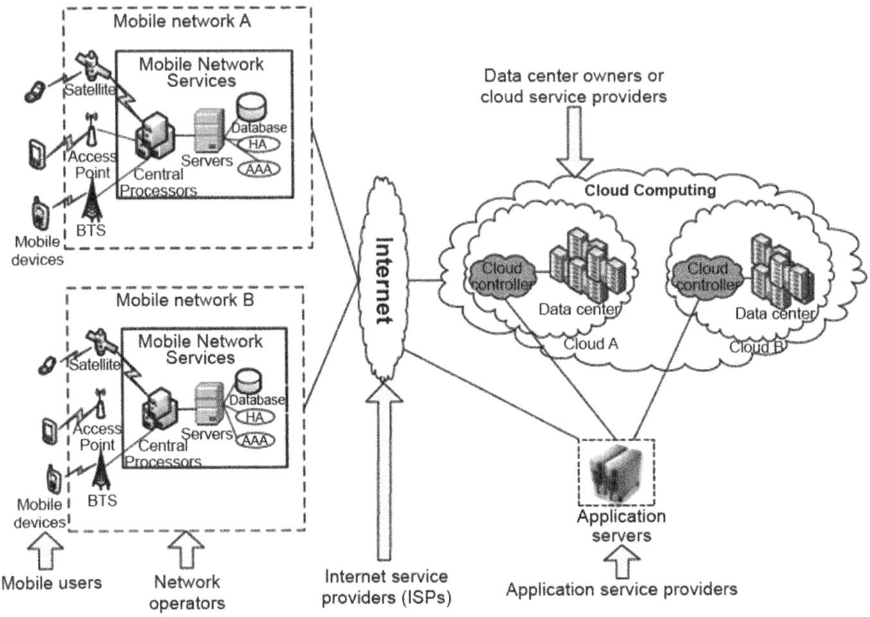

Figure 6.4. *Generic architecture of MCC (Gupta and Gupta 2012)*

6.3.1.1. *Mobile device*

The mobile device is a wireless portable device executing computation-intensive tasks. Smartphones, laptops, personal digital assistants (PDA), wearable devices, sensors and embedded systems (such as RFID readers and biometric readers) are examples of mobile devices in charge of computation-intensive tasks. It is worth noting that mobile devices are, by their nature, characterized by limited battery lifetime and computing power.

6.3.1.2. *Access network*

The role of the radio access network is to transmit the input data of the components of the application to be outsourced. It also yields the final result of cloud computations. The transmission of input data to the cloud uses an ascending connection, while the transmission of the final result of the component processing to the mobile device uses the descending connection of the access network. The radio connection characteristics in terms of throughput, interference and channel stability greatly influence processing time and transmission energy. It is however difficult to determine its characteristics in advance and for the whole transmission duration, given their strong variation in time and space. This is due to the dependence of these parameters and other factors, such as the mobile user positioning with respect to the access point, the surrounding sources of interference, the number of customers in the cell as well as the users' transmission power, etc.

6.3.1.3. *Backhaul and backbone networks*

The backhaul network connects the base station or the point of access to the core network (or backbone network), on which the cloud can be accessed. The backhaul network can be wired or wireless, while the core network is generally wired and Internet based. Given that, after having reached the point of access, the input and output data must be also transported on the backhaul and backbone networks, the quality of service offered by these two networks is among the important parameters influencing offloading efficiency.

6.3.1.4. *Cloud*

The cloud is a powerful remote computing and storage center, accessible through the Internet and usable for data storage or computing. Although the use of cloud storage provisioned by large providers on the Internet, here referred to as "distant cloud", offers advantages in terms of computing power and enormous storage space, data transportation through backhaul and backbone connections adds up transmission time leading to an increase in the overall time required for remote task accomplishment.

To address this problem, the edge cloud concept emerged, according to which the processing and storage resources are brought in the proximity of the mobile user (Magurawalage *et al.* 2015; Mach and Becvar 2017). Although the resources of the edge cloud may not be as rich as those of the distant cloud, the former reduces the offloading delay and improves the processing agility by increasing the end-to-end bandwidth and sharing local resources (Zhang *et al.* 2016; Jiang *et al.* 2019).

A mobile form of the edge cloud, composed of close mobile devices, is also possible in order to take advantage of external resources when the connection to the point of access is inappropriate or inaccessible (Zhang *et al.* 2015). This mobile cloud, which is also known as an "ad hoc cloud" or a "D2D[1] cloud", is beneficial not only because it does not depend on wireless connection availability or quality at the point of access, but also due to the significant transmission throughput accessible among local devices, and to sparing the costs related to the use of radio resources.

1 Device-to-Device.

Obviously, the resources made available by the D2D cloud are well below those offered by the distant cloud or by the edge cloud.

6.3.1.5. *Offloading decision middleware*

The offloading decision middleware is a software element of the MCC architecture in charge with the offloading decision. In terms of physical position, this element can be located on the mobile device, outsourced, at the point of access, on the cloud or at a third point dedicated to the offloading decision.

NOTE.– The decision middleware can be implemented inside or outside the mobile device.

6.3.2. *C-RAN-based architecture*

In this approach, the classical radio network is replaced by C-RAN, enabling among others points the enrichment of the offloading decision by knowledge accessible via the cloud of the radio network.

REFRESHER.– C-RAN architecture is a new network architecture in which baseband and channel processing are outsourced to a centralized base band unit (BBU) located in the cloud. Consequently, the wireless antenna, known as the remote radio head (RRH), acts only as a relay that compresses and connects the signals received from the user equipment to the BBU pool through fronthaul wireless connections.

Given the radio communications-dedicated cloud, the offloading decision execution could also be outsourced to the C-RAN. This would enable the execution of the decision by the proximity servers, which have abundant resources. Moreover, the decision may take advantage of the information collected by the C-RAN, concerning user access network throughout its motion from one point to another (characteristics of the visited networks, etc.). This context knowledge would facilitate a more realistic decision making and would thus optimize the run time and the energy consumed to run the application.

6.4. Offloading decision

This section first presents the various models for the positioning of the offloading decision middleware. Then, it describes the decision variables and the offloading modeling.

6.4.1. *Positioning of the offloading decision middleware*

This section describes the various MCC architectures from the perspective of the offloading decision location. The decision middleware positioning is important, given its impact on the complexity admitted during the development of offloading decision algorithms.

6.4.1.1. *Middleware embedded in the mobile device*

In this architecture, whose representation is similar to the one illustrated in Figure 6.4, the offloading decision middleware is embedded in the mobile device. The main advantage of this approach is that offloading is fully independent from any third party as long as the access network and the cloud are accessible. This architectural approach however requires the use of light decision algorithms that can be run within acceptable times on mobile devices with limited computing power.

6.4.1.2. *Outsourced middleware*

In this second architectural approach, the offloading decision middleware is implemented outside the mobile device, most often in the edge cloud or on a server that is close to the mobile user. In the C-RAN-based architecture, the decision middleware can also be implemented in the BBU pool (Cai *et al.* 2016). Important decision parameters such as radio network and cloud characteristics should therefore be transmitted to the decision middleware.

Two models of outsourced decision-based architecture can be distinguished. The first is a model in which only the offloading decision is assigned to the distant middleware; this decision is, nevertheless, returned to the mobile device to be run by the application. The application components chosen for offloading are then transmitted from the mobile device to the cloud. In the second model, besides the WRG of the application, the program code is also transmitted to the middleware in the form of a virtual machine (VM) or container to enable the launching of the decision reached from the decision middleware.

NOTE.– In the outsourced decision-based architecture, the components of the application can thus be transmitted to the decision middleware. The latter makes it possible to apply the decision from the middleware, which avoids returning the decision to the mobile device prior to starting the execution of the application according to the decision made.

The model of the offloading decision application is not that important: the main advantage of this architecture is the computation power of the external middleware compared to that of the mobile device architecture. This property enables the use of more sophisticated decision algorithms with no decision agility loss. However, outsourcing the decision to a distant server introduces an additional delay in the transmission of decision parameters to the middleware.

6.4.2. *General formulation*

The formulation of the offloading problem depends on the single or multiple contexts of the cloud and of the access network in which the offloading is done. Multi-site offloading can use several clouds such as distant cloud, edge cloud, D2D cloud or a combination of these types of cloud. Moreover, in the context of multi-access networks, several access networks of a macro or micro type are accessible. Based on these terminologies, a distinction can be made between single-access single site, single-access multi-site, multi-access single-site and multi-access multi-site.

In what follows, this section considers the most general case of a multi-access multi-site context and proposes a generic formulation applicable to all four offloading environments.

6.4.2.1. *Offloading decision variables*

Considering the multi-access multi-site context, the definition of the offloading decision variables takes into account the problems to which the decision attempts to find a solution. These problems are follows:

– for each application component, determine if it should be locally or remotely run in order to reach an optimal result ("*how to run*"). Given $C=\{c_1,c_2,...,c_N\}$ (set of components of the application where $N=|C|$), the solution to this problem is described by a set $P=\{p_1,p_2,...,p_N\}$ of size N, where each element p_i is 0 (respectively, 1) if a decision is taken to locally run (respectively, offload) the corresponding component c_i ($p_i \in \{0,1\}$). It is worth noting that the representation of the set P is the same for the coarse-grain offloading, where $N=|C|=|P|=1$;

– for each component to be offloaded, determine the cloud where this component should be executed ("*where to execute*"). Given $M=\{m_1,m_2,...,m_K\}$ the set of cloud sites accessible for computation offloading, where $K=|M|$, the solution to this problem is a set $Q=\{q_1,q_2,...,q_N\}$, in which each element q_i describes the execution cloud chosen for the component c_i of the application ($q_i \in M$). Similarly to the previous problem, the representation of the solution Q is the same for coarse-grain offloading, or $|Q| = 1$;

– for the components to be offloaded, determine the access network through which the input data of these components should be transmitted ("*who to transmit through*"). Given $F = \{f_1, f_2, ..., f_K\}$ the set of points of access that are accessible at the data transmission moment, the solution to this problem is a variable $f \in F$ that describes the access network for offloading the components chosen for execution in the cloud.

To minimize the number of decision variables, sets P and Q can be merged into a single set $X = \{x_1, x_2, ..., x_N\}$, in which each element x_i represents the execution site of each application component irrespective of whether it is on the mobile device or in one of the accessible clouds ($x_i \in \{local, m_1, m_2, ..., m_K\}$). The offloading solution S could be described by a pair (X, f), in which each decision variable x_i represents the site of execution of the *i*th component of the application and the variable *f* signifies the point of access from which the offloading data should be transmitted.

6.4.2.2. *Objective function of the offloading problem*

Independently of the offloading granularity, the objective of the offloading decision is to find the offloading solution that optimizes the application execution cost. Cost is generally expressed in terms of application execution time and energy consumption for the application execution. Nevertheless, other criteria such as price and security are worth being considered. The general objective function of the offloading is modeled as described by equation [6.1]:

$$\arg\min_S \cos t(S) \qquad [6.1]$$

where:

– $S = (X, f)$: offloading solution whose details are provided in section 6.4.2.1;

– $\cos t(S)$: cost function of the offloading solution S.

REFRESHER.– The argument of minimum, denoted by "$\arg\min$ ", is the set of points in which an expression reaches its minimal value. Using mathematical notation, for a function f: X → Y, where Y is a fully ordered set, $\arg\min$ is defined by:

$$\arg\min_x f(x) := \{x \mid \forall y : f(y) \ge f(x)\}$$

The cost function is defined by the weighted sum of the application execution time, on the one hand, and the energy consumption, on the other hand, as described by equation [6.2]:

$$cost(S) = w_t \times \frac{T(S)}{T(S_{local})} + (1 - w_t) \times \frac{E(S)}{E(S_{local})} \qquad [6.2]$$

where:

– $T(S)$ and $E(S)$ are, respectively, the execution time and the energy consumption resulting from solution S;

– $T(S_{local})$ and $E(S_{local})$ are, respectively, the time and the energy consumption required by the fully local execution of the application.

These two latter terms are used for the normalization of time and energy in order to render them summable in equation [6.2]. On the other hand, w_t is a coefficient measuring the relative importance of the application execution time with respect to the energy consumption in the calculation of the total cost of an offloading solution S. The value of this parameter can be fixed depending on user preferences and application needs.

6.4.3. *Modeling of offloading cost*

6.4.3.1. *Fine-grain offloading cost*

The modeling of fine-grain offloading cost requires a distinction between two execution modes of program components: serial execution and parallel execution. In the serial execution mode, the application components are executed one after the other in an appropriate order, considering their granularity levels and input data. In the parallel execution mode, the components whose input data are not interdependent can be executed in parallel. For simplicity reasons, the rest of this chapter focuses on the serial execution mode.

Based on this principle, the cost of fine-grain offloading in terms of execution time is the sum of the time consumed for (1) the execution of the application components according to the chosen execution site and (2) the transmission of input data between two connected components. The offloading time in terms of execution time could therefore be modeled as indicated in equation [6.3]:

$$T(S)=\sum_{c=1}^{N}\sum_{m=0}^{K} y_c^m \times t_{c,m}^{exec} + \sum_{c=1}^{N}\sum_{c'=1}^{N} z_{c,c'}^{m,m'} \times t_{c,c'}^{m,m'} \qquad [6.3]$$

where:

– $t_{c,m}^{exec}$: time for the execution of component c at site m;

– $t_{c,c'}^{m,m'}$: time for the invocation between two components c and c' executed at sites m and m', respectively.

The binary variables y_c^m and $z_{c,c'}^{m,m'}$ are defined depending on the following conditions:

$$y_c^m = \begin{cases} 1, x_c = m \\ 0, x_c \neq m \end{cases} \qquad [6.4]$$

$$z_{c,c'}^{m,m'} = \begin{cases} 0, x_c = x_{c'} \\ 1, x_c \neq x_{c'} \end{cases} \qquad [6.5]$$

In equation [6.3], the time for the execution of component c at site m is defined by:

$$t_{c,m}^{exec} = \frac{wl_c}{ps_m} \qquad [6.6]$$

where:

– wl_c : workload of component c;

– ps_m : CPU processing speed offered by site m.

The time of invocation between two components c and c' executed at sites m and m' is represented by equation [6.7]:

$$t_{c,c'}^{m,m'} = \frac{d_{c,c'}}{bw_{m,m'}} \quad [6.7]$$

with:

– $d_{c,c'}$: amount of data transmitted between components c and c' (represented by $w_{cc'}$ in WRG);

– $bw_{m,m'}$: bandwidth of the network connecting sites m' and m'.

The offloading energy is calculated from the standpoint of the mobile device. Consequently, the energy dedicated to the execution of a component in the cloud is not taken into account when modeling the energy cost. Offloading energy is modeled by equation [6.8]:

$$E(S) = \sum_{c=1}^{N}\sum_{m=0}^{K} y_c^{local} \times e_{c,local}^{exec} + \sum_{c=1}^{N}\sum_{c'=1}^{N} z_{c,c'}^{m,m'} \times e_{c,c'}^{m,m',f} \quad [6.8]$$

with:

– $e_{c,local}^{exec}$: energy dedicated to the execution of component c on the mobile device;

– $e_{c,c'}^{m,m',f}$: energy dedicated to inter-component transmissions.

These two parameters are defined as follows:

$$e_{c,m}^{exec} = t_{c,local}^{exec} \times p_{local}^{cpu} \quad [6.9]$$

$$e_{c,c'}^{m,m',f} = \begin{cases} t_{c,c'}^{m,m'} \times p_{local}^{transmit}, m = local \\ t_{c,c'}^{m,m'} \times p_{local}^{recv}, m' = local \end{cases} \quad [6.10]$$

with:

– p_{local}^{cpu} : CPU energy consumption of the mobile device per processing unit;

– $p_{local}^{trqnsmit}$ and p_{local}^{recv} : transmission power and reception power of the mobile device antenna.

6.4.3.2. *Coarse-grain offloading cost*

Coarse-grain offloading cost could be modeled as a specific fine-grain offloading case with a single component. This is represented as follows:

$$T(S) = \sum_{m=0}^{K} y_c^m \times t_{c,m}^{exec} + z_{main,c}^{local,m} \times t_{main,c}^{local,m} \qquad [6.11]$$

Similarly, the offloading energy cost is modeled as follows:

$$T(S) = y_c^{local} \times e_{c,local}^{exec} + z_{main,c}^{local,m} \times e_{main,c}^{local,m} \qquad [6.12]$$

6.5. AI-based solutions

The previously described fine-grain offloading problem is an NP-hard problem (Wang *et al.* 2015; Wu *et al.* 2016). Therefore, the solution to this problem cannot be found in a polynomial time. Consequently, the optimal solutions and their AI-based variants are applicable only to small scenarios, with a limited number of decision variables. Non-optimal solutions based on AI, such as heuristics or metaheuristics, are used for wider scenarios. Some of the metaheuristics used are worth being mentioned: simulated annealing, tabu search method, evolutionary algorithms, ethology-based algorithms, etc. The following section presents several AI-based optimization algorithms for solving this problem.

6.5.1. *Branch and bound algorithm*

Branch and bound (B&B) is an optimal algorithm for solving combinatorial optimization problems. Several AI-based techniques were developed with the purpose of optimizing the performances of this algorithm in terms of time required for computing the optimal solution.

In the offloading solutions based on this algorithm, the WRG of the application is first transformed into a tree representing all the possible combinations of local or remote execution of each WRG vertex. For this purpose, the tree is built from an empty root. The first component of the application is added to this root in as many copies as the possible execution sites. Each node of this level therefore represents the first component labeled with one of the execution sites. Similarly, at each child thus produced, the second component of WRG is added in as many copies as the possible execution sites. This process is repeated until all the WRG vertices are added to the tree. Consequently, each branch of the tree thus produced represents a potential solution to this offloading problem. The weight of each vertex in WRG is copied in each copy of this component in the tree. However, the connection between a vertex and its parent in the tree takes the weight of the corresponding connection in the WRG, if there is a connection between these two components. Otherwise, it has zero weight. Figure 6.5 illustrates the example of a tree corresponding to a WRG with three vertices in a context with three possible execution sites (local site, Cloud m1 and Cloud m2).

In the evaluation phase, each branch is explored using the depth-first search strategy. When a branch is thus explored, its cost is progressively calculated using the cost function defined by equation [6.2]. In order to accelerate the tree exploration, a branch could be cut when its partial cost exceeds the bounding value. In our offloading decision problem, the bounding value is initiated at the minimum local or remote offloading cost of all the components in one of the cloud sites. When a branch is fully explored without being cut, the bounding value is replaced by the cost of the solution given by this branch.

To accelerate the resolution of the offloading problem, there are solutions based on the B&B algorithm that propose optimizations to enable the earlier cutting of inappropriate branches. For example, the solution proposed by Goudarzi *et al.* (2016) places the vertices with more significant weights higher in the tree, thus enabling inappropriate branches to be cut earlier. On the other hand, it is possible to initiate the bounding value at the cost of the solution found by heuristics or metaheuristics.

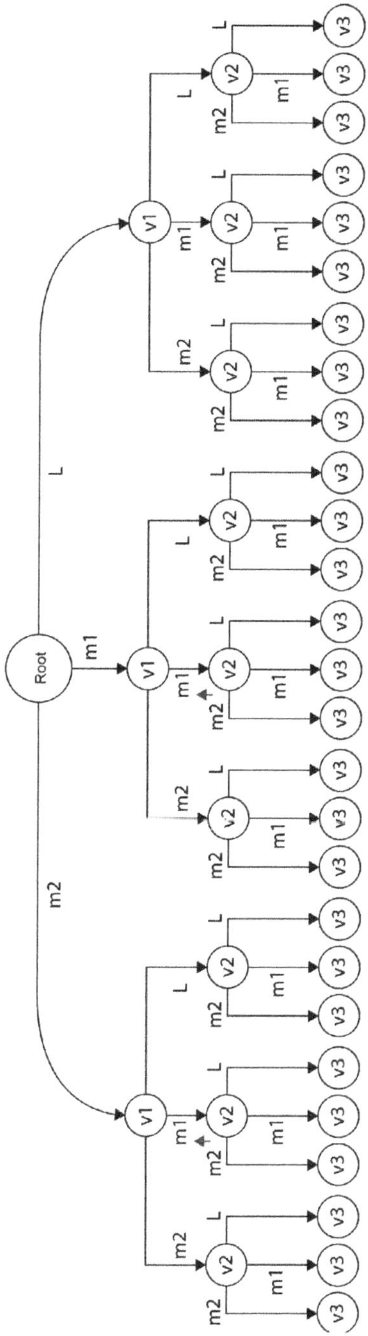

Figure 6.5. *B&B tree*

6.5.2. *Bio-inspired metaheuristics algorithms*

In order to solve the offloading decision problem in a polynomial time, some work relies on evolutionary metaheuristics algorithms, such as the genetic algorithm. The following part of this section describes the general idea of the works relying on this type of solution.

First of all, the initial population is generated from a number of potential solutions. Each solution is described by a chromosome composed of a set of genes. Each gene represents a component of the application labeled by a site of execution. The initial population is generally generated by random approach. Nevertheless, in some of the proposed optimizations, the initial population contains chromosomes with all the genes labeled to be executed in the same site. Figure 6.6 illustrates the chromosome example.

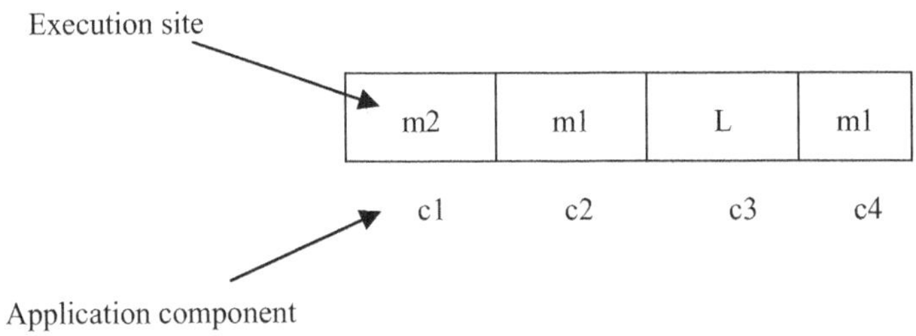

Figure 6.6. *Chromosome*

The algorithm tries to produce a new generation from a given population using selection and reproduction processes. These two processes enable the population optimization in each iteration. The selection process generally relies on the fitness function, which measures the relevance degree of a solution to the aimed objective. For our offloading decision problem, the fitness function corresponds to the cost function given by equation [6.2], which is expected to be minimized. Consequently, the selection stage relies on a probability that is inversely proportional to the cost calculated for each chromosome from equation [6.2].

After the selection phase, mutation and crossover functions are applied to the chromosomes selected for the reproduction of a new generation. In most works based on genetic algorithms, parent crossover relies on methods such as cutting and exchange. On the other hand, genes are chosen for mutation by roulette wheel selection function. The latter assigns a cost (fitness) to each gene of a chromosome.

The fitness of a gene is calculated as the cost of the gene divided by the cost of its chromosome. Consequently, the gene with the highest cost has the highest probability to be chosen for mutation. The selected gene is mutated by randomly changing its assigned site.

Selection, mutation and crossover stages unfold until (1) the maximal number of iterations is reached or (2) no significant improvement is observed when passing from one generation to the next.

6.5.3. *Ethology-based metaheuristics algorithms*

Some work focuses on an offloading decision relying on ethology-based metaheuristics, such as ant-colony optimization (ACO), particle swarm optimization (PSO) and artificial bee colony optimization (ABC). The following part of this section focuses on the ACO-based offloading decision, as a typical example of ethology-based metaheuristics.

ACO is a metaheuristics algorithm inspired by the collective behavior of ants when they try to find the best path to their nest and a food source. In nature, an ant starts by randomly looking for food. Once it finds food, the ant returns to the nest leaving a trail of a chemical substance known as pheromone. Because of this trail, other ants are able to find the food. In time, the pheromone trail starts to evaporate. The longer it takes for an ant to travel along the path, the stronger the evaporation of pheromones. By comparison, a short path is more often traveled up and down by ants and the pheromone density is higher on shorter paths than on longer paths.

Relying on this principle, some work represents each potential solution to the offloading problem by the path traveled by the ant. Initially, a number of ants are randomly generated. With each algorithm iteration, each ant updates its path depending on the relevance of the execution site chosen for each component in its previous path. For our problem, the latter is defined by a fitness function given by equation [6.2]. These stages unfold until a maximal number of iterations are reached.

6.6. Conclusion

This chapter presents the MCC paradigm, which extends the capacities of mobile devices taking advantage of abundant cloud resources. The notion of computation offloading is defined in this context. This involves the outsourcing of complicated computations to the cloud. Two offloading types are introduced, namely the fine-grain and the coarse-grain offloading, reflecting the full or subpart application offloading. Given the variable conditions of radio channel and cloud site, the necessity of an offloading decision is highlighted. In this context, the offloading decision mechanism should determine which cloud and access network should be preferred for computation offloading. This multi-site multi-access environment leads to an NP-hard problem, which cannot be solved in a polynomial time. To solve this problem, the literature proposes AI-based solutions, including heuristics and metaheuristics algorithms. To clarify the approaches taken to solve this problem, various classes of applied algorithms are presented.

Although the general offloading problem described in this chapter is thoroughly studied in the literature, there are no in-depth studies of problems related to dynamic and mobile context, multi-clouds and multi-users. In particular, the dynamic and mobile context requires the introduction of very light decision algorithms, which can be reactively executed following context changes. Certain works along this axis rely on user mobility prediction and deal also with shifted offloading. On the other hand, determining the price of radio resources and cloud is a very interesting domain for which solutions based on game theory and auction theory should be proposed. A further research direction involves proposing algorithms for the allocation of edge cloud and access network resources according to the needs of all users. Joint allocation of radio and cloud resources enables significant optimization of the quality of service, user admission rate and use of resources.

6.7. References

Abolfazli, S., Sanaei, Z., Ahmed, E., Gani, A., and Buyya, R. (2014). Cloud-based augmentation for mobile devices: Motivation, taxonomies, and open challenges. *IEEE Communications Surveys & Tutorials*, 16(1), 337–368.

Cai, Y., Yu, F.R., and Bu, S. (2016). Dynamic operations of cloud radio access networks (C-RAN) for mobile cloud computing systems. *IEEE Transactions on Vehicular Technology*, 65(3), 1536–1548.

Chen, M., Hao, Y., Li, Y., Lai, C.F., and Wu, D. (2015). On the computation offloading at ad hoc cloudlet: Architecture and service modes. *IEEE Communications Magazine*, 53(6), 18–24.

Goudarzi, M., Movahedi, Z., and Pujolle, G. (2016). A priority-based fast optimal computation offloading planner for mobile cloud computing. *International Journal of Information & Communication Technology Research*, 8(1), 43–49.

Gupta, P. and Gupta, S. (2012). Mobile cloud computing: The future of cloud. *International Journal of Advanced Research in Electrical, Electronics and Instrumentation Engineering*, 1(3), 134–145.

Hong, S. and Kim, H. (2019). QoE-aware computation offloading to capture energy-latency-pricing tradeoff in mobile clouds. *IEEE Transactions on Mobile Computing*, 18(9), 2174–2189.

Jiang, C., Cheng, X., Gao, H., Zhou, X., and Wan, J. (2019). Toward computation offloading in edge computing: A survey. *IEEE Access*, 7(1), 131543–131558.

Khan, A.U.R., Othman, M., Madani, S.A., and Khan, S.U. (2014). A survey of mobile cloud computing application models. *IEEE Communications Surveys & Tutorials*, 16(1), 393–413.

Khan, A.U.R., Othman, M., Xia, F., and Khan, A.N. (2015). Context-aware mobile cloud computing and its challenges. *IEEE Cloud Computing*, 2(3), 42–49.

Li, B., Liu, Z., Pei, Y., and Wu, H. (2014). Mobility prediction based opportunistic computational offloading for mobile device cloud. *IEEE 17th International Conference on Computational Science and Engineering (CSE)*. IEEE, Chengdu, 786–792.

Mach, P. and Becvar, Z. (2017). Mobile edge computing: A survey on architecture and computation offloading. *IEEE Communications Surveys & Tutorials*, 19(3), 1628–1656.

Magurawalage, C.S., Yang, K., and Wang, K. (2015). Aqua computing: Coupling computing and communications [Online]. Available at: https://arxiv. org/abs/1510.07250, [Accessed January 2019].

Movahedi, Z. (2018). Green, trust and computation offloading perspectives to optimize network management and mobile services. Doctoral Thesis, Sorbonne Universities, Paris.

Wang, X., Wang, J., Wang, X., and Chen, X. (2015). Energy and delay tradeoff for application offloading in mobile cloud computing. *IEEE Systems Journal*, 11(2), 858–867.

Wu, H. (2018). Multi-objective decision-making for mobile cloud offloading: A survey. *IEEE Access*, 6(1), 3962–3976.

Wu, H., Knottenbelt, Z., Wolter, K., and Sun, Y. (2016). An optimal offloading partitioning algorithm in mobile cloud computing. *Quantitative Evaluation of Systems*, Agha, G. and Van Houdt, B. (eds). Springer, Berlin, 311–328.

Zhang, Y., Niyato, D., and Wang, P. (2015). Offloading in mobile cloudlet systems with intermittent connectivity. *IEEE Transactions on Mobile Computing*, 14(12), 2516–2529.

Zhang, K., Mao, Y., Leng, S., Zhao, Q., Li, L., Peng, X., Pan, L., Maharjan, S., and Zhang, Y. (2016). Energy-efficient offloading for mobile edge computing in 5g heterogeneous networks. *IEEE Access*, 4(1), 5896–5907.

PART 4

AI and New Communication Architectures

7

Intelligent Management of Resources in a Smart Grid-Cloud for Better Energy Efficiency

Mohammed Anis BENBLIDIA[1], Leila MERGHEM-BOULAHIA[1], Moez ESSEGHIR[1] and Bouziane BRIK[2]

[1] *University of Technology of Troyes, France*

[2] *CESI, Rouen, France*

7.1. Introduction

With the boom of the Internet of Things (IoT), our world is becoming increasingly connected. According to a study conducted by Gartner and the European Audiovisual and Telecommunications Institute (*Institut de l'audiovisuel et des télécommunications en Europe* [IDATE]), the number of connected objects in the world will reach 50 billion by 2030[1]. Given that connected objects generally require cloud computing and storage services, cloud data centers will need to process a considerable number of user queries.

Consequently, the major role played by cloud data centers in the storage, computing and management of data generated by the IoT will exponentially grow over time. One of the greatest challenges in the development of cloud data centers is

1 Available at: https://www.juniperresearch.com/press/press-releases.

Intelligent Network Management and Control,
coordinated by Badr BENMAMMAR.

their high energy consumption. Moreover, the data centers sector is estimated to account for 1.4 % of global electricity consumption (Avgerinou *et al.* 2017).

On the other hand, the smart grid – which is the future intelligent electrical grid – enables a bidirectional exchange of data and energy between producer and consumer. Because of this data exchange, the smart grid administrator has more information on its clients and is able to offer them more customized services that further meet their needs.

This work is a closer examination of the energy efficiency of information and communication infrastructures in a smart grid-cloud. It deals in particular with communication networks and cloud data centers. Our focus is on the latter, due to their high consumption of energy, which confers them an important role in the network. The rest of the chapter is organized as follows. First, section 7.2 gives an overview on smart grids and their interactions with cloud data centers. Section 7.3 highlights a state-of-the-art on various energy efficiency techniques used in data centers. Section 7.4 introduces decision-aiding techniques in a smart grid-cloud. Section 7.5 concludes the chapter.

7.2. Smart grid and cloud data center: fundamental concepts and architecture

Although almost unchanged for about a century, the electrical power grid infrastructure as we know it today has successfully met our needs. But since it is ageing in time, this infrastructure is becoming less and less efficient, is continuously confronted with its limits and permanently struggling to address our demands. Moreover, this electrical power grid generates electricity in a centralized manner and distributes it to a large number of users through one-way communication: from producers to consumers. The need for reliability, easy energy management and renewable energy production emphasizes the necessity of an upgraded and smart grid in the future. In this context, the smart grid opens the path for the future electrical power grid, which aims to create a clean, safe, secure and reliable system (Markovic *et al.* 2013).

The management of the electrical power grid requires several significant changes without disrupting its operation. In the context of the electrical power grid, this is reflected by the set-up of a network of sensors and smart counters that forward production and consumption data to the smart grid administrator. The latter then starts communication and interaction.

7.2.1. *Network architecture for smart grids*

The smart grid can be considered as a network of many systems and subsystems that are smartly interconnected to efficiently and reliably provide energy. Smart grid intelligence is enhanced by adding information and communication technologies (ICT) to the electrical power grid (Gungor *et al.* 2011). This communication infrastructure enables gathering data on production and transportation as well as on the distribution networks. In terms of communication infrastructure, the smart grid can be divided into three main entities: home area network (HAN), neighborhood area network (NAN) and wide area network (WAN) (Hossain *et al.* 2012).

7.2.1.1. *Home area network*

HAN is the smallest subsystem in the hierarchical chain of the smart grid. It represents a single residential unit featuring smart devices, energy consumption control devices, storage solutions, solar panels, small wind turbines, electrical vehicles (EV) and smart meters (SM). These components intercommunicate using wired communication technologies such as power line communication (PLC) and wireless technologies such as Wi-Fi and ZigBee. Unlike wired technologies, wireless technologies, such as ZigBee, are becoming a popular choice due to their low installation cost and their flexibility (Yan *et al.* 2013).

7.2.1.2. *Neighborhood area network*

NAN is a set of HAN networks, which corresponds to group of houses potentially supplied by the same transformer. NAN implements the connection that makes it possible to offer several smart grid applications such as smart metering, load management, energy distribution and power failure management automation. Advanced metering infrastructure (AMI) collects data from NAN smart devices and groups them before they are sent to WAN. The latter makes the connection between NAN and the power grid company.

7.2.1.3. *Wide area network*

WAN connects several distribution systems and serves as a bridge between NAN and HAN and the company grid. As shown in Figure 7.1, WAN provides the communication infrastructure required for connecting the devices of smart grid clients to the power grid company. This can be done by means of several communication technologies (Ethernet, cellular networks, etc.) for the transfer of data issued by NAN to the grid company.

Relevant examples of the main applications offered by WAN are long distance/remote monitoring, control and protection. These applications are efficient solutions for the improvement of planning, operation and protection of the electrical power network in the smart grid (Terzija *et al.* 2011). Monitoring, control and protection applications in a WAN offer higher data resolution and shorter response time than the classical supervisory control and data acquisition (SCADA) systems and energy management systems (EMS). The latter offer a measurement update interval of several seconds, or several minutes, while monitoring, control and protection applications on the WAN provide high-resolution data, namely 60 samples per second (Khan *et al.* 2016).

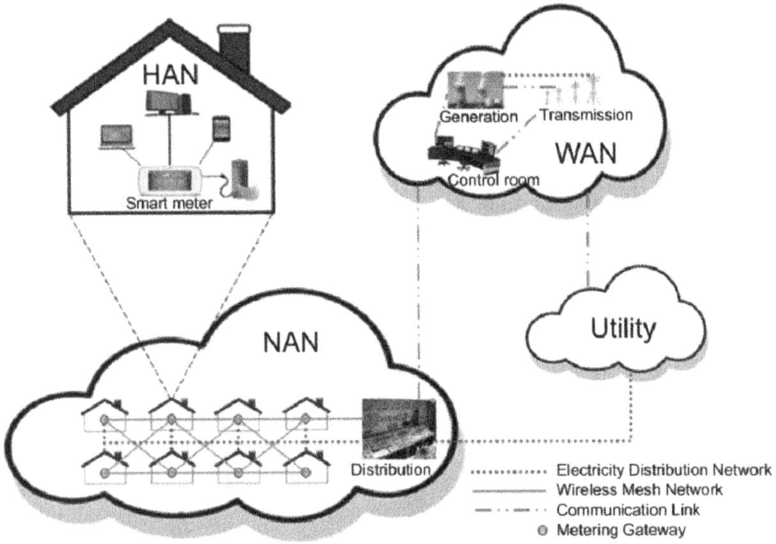

Figure 7.1. *Smart grid network architecture*[2]*. For a color version of this figure, see www.iste.co.uk/benmammar/network.zip*

7.2.2. *Main characteristics of smart grids*

The smart grid supplies its clients with energy by managing mainly three technical hubs: generation, transportation and distribution. The generation hub involves the traditional electrical power plants. The transportation hub ensures the delivery of the generated electricity to the distribution hub. The latter is in charge of distributing the energy to the smart grid clients. These three main components are illustrated in Figure 7.2.

2 See Bera *et al.* (2015).

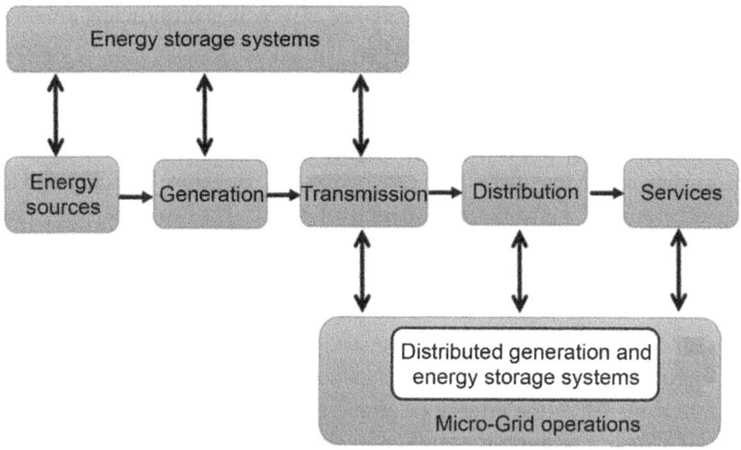

Figure 7.2. *Generation, transportation and distribution systems in the smart grid*

One of the major objectives of the smart grid is the control of client energy consumption by the implementation of various optimization methods. To reach this objective, the smart grid uses mainly the following: smart metering (bidirectional communication) and the microgrids (MG; distributed energy resources [DER]), demand response (DR) mechanisms and smart pricing (Bera *et al.* 2015).

7.2.2.1. *Bidirectional communication*

To create an automated and widely distributed network, the smart grid adds to a traditional electrical network the advantages of communication technologies, particularly sensors, connected objects, and SM to provide real-time information and enable a quasi-instantaneous balance of supply and demand management. This is why it is important to implement a reliable communication infrastructure enabling robust real-time data transmission via extended networks (WAN) to the client (Bera *et al.* 2015).

7.2.2.2. *Distributed energy resources*

A smart grid enables the integration of DER involving several technologies: fuel cells, photovoltaic panels, wind turbines, etc. DER coordinated control, as well as the controllable loads of storage devices, such as flywheels and batteries, are at the core of the MG concept.

MG is an electrical network composed of energy consumption sources, (renewable or thermal) electrical power generation systems, energy storage systems and data management and control systems (Guerassimof 2017). It is an efficient solution to the problems raised by traditional infrastructures. The capacity of the MG to integrate local generators, as energy generation resources, and storage systems offers many advantages compared to existing systems. MG scan operate while being connected to the main electrical distribution network, where their storage systems enable a local optimization of supply and demand. Moreover, they can operate in isolation in island mode, being disconnected from the network for a period of time[3]. The two operation modes are managed by the operator depending on the technical or economic situation of the electrical system.

One of the advantages of integrating MG in the smart grid architecture is the increase in reliability. This is done by ensuring emergency power supply during network failure or when the energy price is on the rise. Moreover, MGs enable the reduction of CO_2 emissions by supplying clients from renewable energy sources (Asmus 2017).

7.2.2.3. *Management of the electrical system equilibrium through DR mechanisms*

DR is an approach used for reducing electrical network load and increasing system reliability. According to the DR approach, the final users modify their electricity consumption models depending on price variations. From smart grid perspective, the DR is an efficient means to schedule user energy consumption in order to reduce the operating expenses due to costly generators. Moreover, due to this technology, the electrical system becomes more reliable, the electrical market transparency and efficiency improve and mutual financial advantages are generated for the smart grid operator and its clients.

The DR mechanism can be implemented by three methods (Deng *et al.* 2015):

– reduce the peak energy consumption in order to prevent the electrical overload of the network supplying the distribution substations. Users' satisfaction would decrease, as this method partly reduces their demand;

– promote off-peak energy consumption by means of energy storage devices, rechargeable batteries and EV;

– balance energy consumption over a time horizon. For example, transfer a part of the electrical load from peak hours to off-peak hours.

3 See http://www.smartgrids-cre.fr.

The DR concept was used in the commercial and industrial fields for a certain time in order to improve network stability. Due to the emergence of smart electrical grids, DR has now the potential to be extended at large scale on the residential electricity markets (Yan *et al.* 2013).

7.2.2.4. *Smart pricing*

Smart pricing is an alternative to fixed electricity prices. Price-based programs are mainly integrated in the DR programs in the smart grid. They offer users different electricity prices at different moments. Based on this information, the users naturally consume less electricity when the prices of electricity are high and therefore reduce the peak hour demand. In other words, these programs indirectly encourage users to dynamically modify their energy use diagrams depending on energy use variation, instead of directly controlling their loads. There are various pricing programs depending on the time scale (Deng *et al.* 2015):

– *time of use (ToU)*: the most commonly used form of dynamic pricing. According to this pricing mode, the day is divided into time blocks, each with a specific price. When the users consume energy at different moments of the day or during different seasons of the year, they are billed at different prices. To encourage users to shift their loads during the day, the announced peak hour electricity price is higher than the one for off-peak periods. In this type of pricing, prices are often fixed and announced in advance and they remain unchanged for a long period of time (Vardakas *et al.* 2015);

– *critical peak pricing (CPP)*: mainly based on ToU pricing, except for certain days. According to CPP, an emergency signal is sent one day in advance to announce a critical peak period (Bergaentzlé and Clastres 2013). During this peak period, the price increases significantly compared to the usual price, strongly encouraging users to avoid consumption at that moment. CPP is employed only for a limited number of hours or days per year to ensure system reliability and balance supply and demand;

– *real-time pricing (RTP)*: involves dynamic pricing, where the price of electricity generally varies at different time intervals throughout the day (every 15 minutes or every hour) (Allcott 2009). RTP is the most efficient pricing, as it enables the consumer to be hourly informed on the kilowatt-hour price on the wholesale market. It gives the real cost of electricity at the moment when it is consumed;

– *inclining block rate (IBR)*: a progressive pricing aimed at reducing the global electricity consumption. It involves a two-level pricing structure (lower and higher blocks), so that the electricity unit price to be paid by the consumer increases with the amount of energy consumed. In other terms, the price of electricity per energy consumption reaches a high value if the user's hourly, daily or monthly energy consumption exceeds a certain threshold. IBR encourages users to divide their loads between different moments of the day to avoid the highest rates, which contributes to reducing the network peak-to-average rate (Gabr *et al.* 2018).

7.2.3. *Interaction of cloud data centers with smart grids*

Victims of their own development, cloud computing and smart grid are faced with several challenges. In particular, the annual electricity invoices of large providers of cloud computing services are increasing every year. As an example, this invoice reached 67 million dollars at the end of 2014 (Deng *et al.* 2014), and the figure is increasingly higher with the expansion of cloud computing services and the rise in electricity prices. In parallel, the smart grid integrates a large number of DER, such as solar panels and wind turbines, and must also provide high operational stability. This may lead to economic difficulties, given the intermittent nature of the decentralized energy production.

The above-mentioned concerns related to cloud computing and smart grids can be addressed by appropriate cooperation between the two parties. The DR of data centers can be a strong asset for the smart grid, as it is motivated by user demands that can be assigned to geographically distributed data centers and serviced by several energy sources. The energy consumed by a data center from the smart grid can be flexibly adjusted by balancing the workload or by modulating the on-site electricity generation. In this context, the adoption of storage means by the data centers in the smart grid makes it possible to address many problems and is of interest to many various concerned parties: producers, consumers and network managers. The energy storage systems may ensure the network equilibrium and stability while improving the efficiency of the transmission and distribution facilities. They represent solutions that play an essential role, in economic and environmental terms, ensuring efficient and sustainable provision of electrical energy (Guerassimof and Maizi 2013).

The interaction between cloud data centers and the smart grid administrator is done directly, through data exchange via the communication infrastructure of the smart grid. But it is also done indirectly via the queries issued by the users to the cloud data centers. Figure 7.3 illustrates the interaction between the cloud data centers and the smart grid and its users.

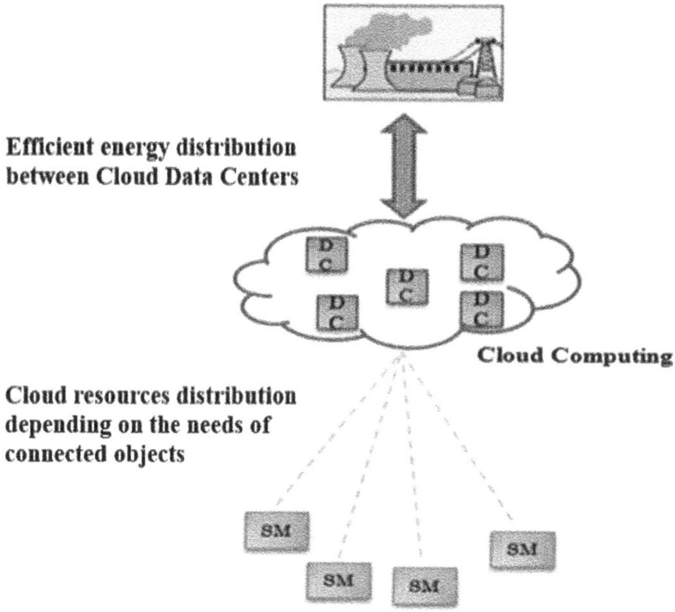

Figure 7.3. *Interaction of data centers with the smart grid and its users*

In a smart grid-cloud system, the user queries are grouped by a cloud aggregator. The latter is in charge of the distribution of queries to the cloud data centers according to a predefined policy. One of the greatest challenges to be addressed is the execution of these queries while ensuring a satisfactory quality of service. In this context, the various queries issued by connected objects (IoT) require intensive processing and low latency, particularly for real-time applications. It is very difficult for cloud computing to cope with this increasingly restricting communication and meet all the demands. Fog computing is an emerging extension of cloud computing architecture, which enables data processing as close as possible to their source and provides the network connections required to forward data from the fog node to the user terminal. Each fog node has processing capacities that enable it to respond in the shortest time to the queries of objects that are connected to the network. Therefore, in the presence of several fog nodes close to the endpoints of the terminal, better performances can be obtained in terms of delay than with cloud computing (Chiang and Zhang 2016; Chiang *et al.* 2017). From this perspective, fog computing does not aim to replace cloud computing, but to complete it in the new fog–cloud computing paradigm, which aims to satisfy the user applications (Elmroth *et al.* 2017). Figure 7.4 illustrates a fog–cloud computing system.

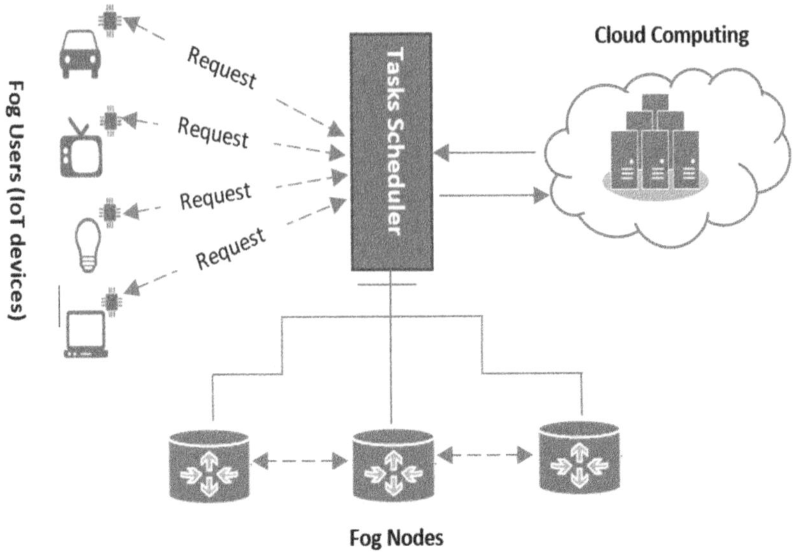

Figure 7.4. *Fog–cloud computing system*

7.3. State-of-the-art on the energy efficiency techniques of cloud data centers

The improvement of data center energy efficiency in terms of cost, consumption or environmental impact has become a subject attracting many researchers. Many research works consider the problem of high energy consumption of data centers and propose solutions implementing various techniques. This section reviews the efforts conducted on this subject and is structured into three subsections: energy efficiency techniques of installations except for information technology (IT) equipment, energy efficiency techniques of data center servers and, finally, energy efficiency techniques of a set of data centers.

7.3.1. *Energy efficiency techniques of non-IT equipment of a data center*

A data center has several installations: servers, inverters, cooling equipment, an operation hall, etc. The major part of the energy consumed by a data center is used for cooling servers and inverters. The basic rule for the management of data center installations is to appropriately organize hot air islands and cold air islands. A cold air island is formed by the intake of servers at the front, while a hot air island is formed by the outlet of hot air from server rows. Ground elevation and the use of cooling equipment at higher temperatures are among the energy saving techniques. To reduce the energy consumption of data centers containers, Endo *et al.* (2013) combined direct cooling using cool air with evaporative cooling in order to reduce the heat released by the IT equipment of the data center. Somani *et al.* (2009) presented an algorithm referred to as "ambient intelligence-based load management", which improves the heat dissipation capacity of the data center. The algorithm uses the input temperature of the server racks and distributes the workloads depending on the thermal environment of the data center. To analyze the efficiency of the implemented techniques, computational fluid dynamics (CFD) is used for modeling the air flow and the heat transfer in the data centers. Joshi and Kumar (2012) classified the CFD models of data centers into two categories: air flow modeling enabling the prediction of air flow rates in the perforated tiles and modeling of the thermal effects of rack display and energy distribution. They also presented alternative diagrams of supply and return of several thermal performance metrics. The energy efficiency of the installations of a data center is generally independent of the smart grid concepts.

7.3.2. *Energy efficiency techniques in data center servers*

Considering the availability of a set of data centers, this category improves the energy consumption of each data center individually, by means of several methods, for example, minimizing the use of their computing equipment, such as processors. The energy efficiency of the computing equipment in a data center is implemented via an energy material, as well as by the virtualization and dynamic consolidation of workloads (Cavdar and Alagoz 2012). Although data centers contain intra-data center network switches, the total consumption of central switches, aggregation and access represents around half of the electrical consumption of IT equipment (Kliazovich *et al.* 2010). Consequently, most efforts on this subject mentioned in the literature relate to the energy efficiency of the servers of a data center.

Gandhi *et al.* (2009) propose an algorithm that makes it possible to find an optimal allocation of electricity to the servers of a data center in order to minimize their average response time. For this purpose, they use two mechanisms for processor voltage and frequency rescaling: *dynamic frequency scaling* (DFS) and *dynamic voltage and frequency scaling* (DVFS). The latter are among the most commonly used techniques for reducing the energy consumption of servers. Wang *et al.* (2013) propose an application that enables the virtualization of electricity distribution in a data center. Their proposal enables the servers to define their needs in terms of energy and, consequently, the application distributes the electricity in a fair manner. Liu *et al.* (2018) consider various sources of energy and propose an architecture for energy management referred to as Datacenter Inner Power Switch Network (DiPSN). The latter efficiently connects various energy sources to appropriate servers.

Resources allocation and migration of virtual machines (VM) considering energy optimization were widely taken into account. Dai *et al.* (2016) modeled the positioning of VM in the servers of a data center as a linear optimization problem. The authors developed two approximation algorithms known as "MinES" and "MinCS" in order to obtain the optimal solution that minimizes energy consumption and ensures the user service level. Sharma *et al.* (2019) propose a hybrid approach that combines genetic algorithms (GA) and particle swarm optimization (PSO) known as "HGAPSO". This algorithm enables the migration of VM while reducing the energy consumption and avoiding the breach of predefined service level. In another work, Duong-Ba *et al.* (2018) deal with the problem of positioning and optimal migration of VM in order to minimize the use of resources and the energy consumption of cloud data centers. They modeled the problem as a multiobjective function and solved it by proposing two multilevel algorithms, which combine positioning and migration of VM.

7.3.3. *Energy efficiency techniques for a set of data centers*

This section considers the techniques designed for a set of data centers geographically distributed similarly to those owned and used by Google, Yahoo, Microsoft, Amazon, eBay, etc. The general idea is to direct the user queries to one or several data centers in order to minimize their cost, their consumption and their CO_2 emission.

This category takes advantage of the functionalities offered by the smart grid to improve the energy efficiency of data centers in the cloud. Moreover, we examine the adoption of data centers for hosting smart grid services in order to have a complete description of the interactions between the smart grid and cloud data centers. The focus is on energy efficiency techniques for several data centers and their corresponding transportation networks. These techniques are applied under two architectures: smart grid-cloud architecture and smart micro-grid-cloud architecture.

7.3.3.1. *Smart grid-cloud architecture*

Several research works considered the interaction of cloud data centers with the smart grid for the functionalities offered by the latter, such as dynamic pricing and DR. Gu *et al.* (2016) considered a smart grid-cloud system where cloud data centers are supplied by the smart grid and by the renewable energy that it self-generates. They propose a user query distribution algorithm that minimizes the energy costs of cloud data centers. This work was extended in Gu *et al.* (2018) by proposing a green scheduler that minimizes both the energy cost and the carbon emissions. In another work, Wang *et al.* (2016a) modeled the interaction between electrical networks and cloud data centers as a two-stage problem. They first use dynamic pricing to achieve load balancing, and then data centers react to the energy price and manage the user query distribution to minimize their total energy cost. Wang and Ye (2016) propose a DR algorithm that considers renewable energies in order to minimize energy costs. Their proposal jointly optimizes the energy supply and the allocation of workload issued by the cloud data center users. Kiani *et al.* (2018) aim to maximize profits associated with the use of renewable energy in geographically distributed data centers. In their system, they consider both the generation of renewable energy and the dynamic pricing of electricity markets. Ding *et al.* (2018) propose a stochastic algorithm for resource distribution in order to optimize the energy costs and the CO_2 emissions of data centers.

7.3.3.2. *Smart micro-grid-cloud architecture*

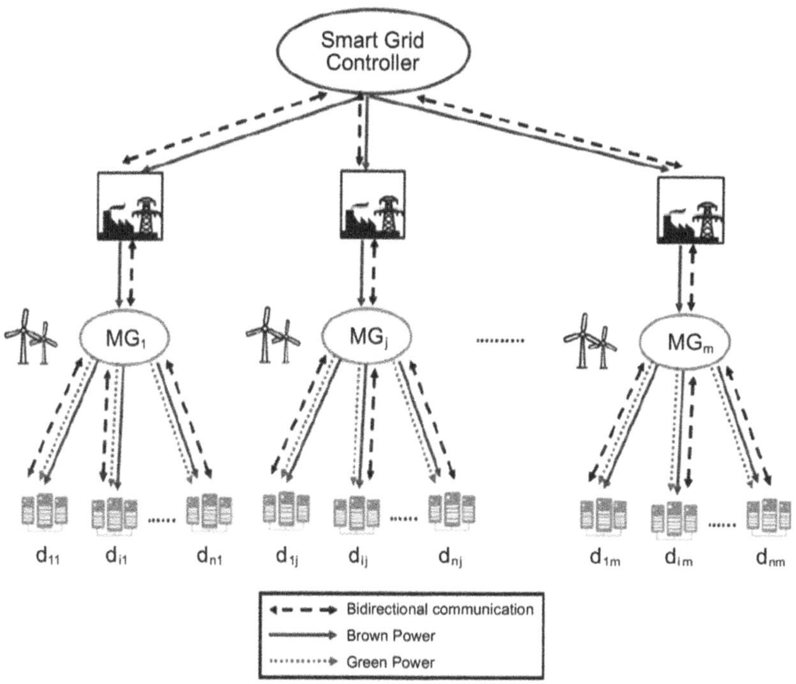

Figure 7.5. *Example of smart micro-grid-cloud architecture. For a color version of this figure, see www.iste.co.uk/benmammar/network.zip*

Considering the interaction of cloud data centers with a smart grid operator yielded significant results in terms of energy efficiency. However, certain studies estimate that running data centers in micro-grids may be more interesting in terms of energy efficiency, given that micro-grids can improve the durability and reliability of electrical services. Figure 7.5 illustrates an example of smart micro-grid-cloud architecture. Yu *et al.* (2016) studied the energy management in cloud data centers supplied by micro-grids. They minimized the energy cost by considering battery charging and discharging efficiency parameters. They also studied in Yu *et al.* (2014) the minimization of energy costs and carbon emissions in data centers supplied by micro-grids. For this purpose, they modeled a stochastic optimization problem and solved it using Lyapunov optimization method. Yu *et al.* (2015) similarly modeled the minimization of data centers energy cost by considering the power failures.

7.3.4. *Discussion*

This state-of-the-art study shows that there are few research studies dedicated to the energy efficiency of IT equipment in a data center from the perspective of the smart grid. Most works dealing with this subject focus on the energy efficiency of a single IT element, without taking advantage of the functionalities brought by the smart grid.

This work takes interest in the techniques considering a set of data centers interacting with the smart grid. Table 7.1 presents a comparative study of the abovementioned works belonging to this category. The comparison relies on four criteria: energy cost reduction, greenhouse gas emissions reduction, electrical network load reduction and integration of renewable energies. It is worth noting that most of the solutions proposed in a smart grid-cloud system are applied on the side of cloud services providers and that the consideration of prices and the use of renewable energies are main factors of energy efficiency in data centers. However, few works examined the role of smart grids in energy management and allocation to cloud data centers. Moreover, the proposed systems do not deal with the influence that the huge energy consumption in data centers may have on the smart grid.

Generally, the objective of research works using micro-grid-based architectures is to reduce the energy cost and the carbon emission of cloud data centers. However, such works do not present a detailed analysis of how their approaches may affect the electrical network, as they ignore the fact that a smart grid provides a limited amount of energy.

Research works	Architecture		Considered metrics			
	Smart grid-cloud	Smart mico-grid-cloud	Energy cost	Carbon emissions	Electrical charge on the network	Renewable energy resources
(Gu *et al.* 2016, 2018)		X	X	X		X
(Wang *et al.* 2016)		X	X		X	
(Wang and Ye 2016)		X	X			X
(Kiani and Ansari 2018)		X	X	X		X
(Ding *et al.* 2018)		X	X	X		X
(Yu *et al.* 2016)	X		X			X

Table 7.1. *Comparative table of works*

7.4. State-of-the-art on the decision-aiding techniques in a smart grid-cloud system

Energy efficiency techniques in a smart grid-cloud system require the interaction of data centers with the smart grid operator. Each of these two entities has a behavior toward the other. These behaviors can be modeled using utility functions and cost functions.

Utility function: the behaviors of various data centers are modeled by various choices of the utility functions. In more formal terms, the utility represents the comfort/satisfaction level obtained by the data center depending on its energy consumption, nondecreasing and concave. Generally, the quadratic utility functions are the most commonly considered (Samadi *et al.* 2012).

Cost function: the cost of electricity generation and provision by the smart grid operator is modeled by the cost function, which is increasing and strictly convex. Two alternatives can be adopted: the piecewise linear cost function (corresponding to IBR) and the quadratic cost function (Mohsenian-Rad *et al.* 2010).

The energy efficiency problems in a smart grid-cloud system are often formulated as optimization problems. The latter can be solved in several ways. This section presents the main approaches to modeling and solving energy efficiency problems in cloud data centers.

7.4.1. *Game theory*

Game theory is a model of interactive decision processes that studies the egoistic or rational behavior of individuals. The latter are referred to as "players" and they are the most important element of the game. A player may represent a single individual or a group of individuals making a decision. After having defined the set of players, two types of models can be distinguished: games whose basic elements are actions of individual players and those based on joint actions of a group of players. The first type are known as "non-cooperative games", while the second type are known as "cooperative games" (Yildizoglu 2011). A game G is composed of three fundamental elements: players *N*; strategies $\{S_i\}_{i \in N}$ and the gain function $\{P_i\}_{i \in N}$. Each player chooses a strategy $s_i \in S_i$ to maximize its gain function $P_i(s_i, s_{-i})$, which depends not only on its strategy s_i, but also on the strategies of the other players s_{-i}.

One of the most important concepts in game theory is the *Nash equilibrium*. It represents the stable and static strategy in which no player has an incentive to unilaterally deviate from its own strategy, given the strategies adopted by the other players. The approaches based on game theory for energy optimization in data centers generally converge toward Nash equilibrium; this leads us to an optimal solution both for supply and demand. Therefore, game theory is an efficient approach for intelligent decision making in a smart grid-cloud system.

In this context, Wang *et al.* (2014a) propose a two-level Stackelberg game between the smart grid controller and the cloud computing controller. The smart grid controller aims to reach load balancing in electrical buses relying on a defined pricing policy. On the other hand, the objective of the cloud controller is to maximize the total price obtained by meeting the user demands. An extension of this work is presented in Wang *et al.* (2014b), where the authors consider an existing electrical network structure (IEEE 24 bus) and the integration of renewable energy sources in the system. Consequently, the load on each supply bus depends both on energy consumption and on renewable energy generation. Benblidia *et al.* (2018) model the interaction between cloud data centers and the smart grid in a non-ccoperative game for efficient energy distribution among data centers. The gain function of the game is modeled in a linear optimization problem and solved via Lagrange multipliers.

7.4.2. *Convex optimization*

The objective functions considered in convex optimization are convex. Generally, energy efficiency methods in data centers aim to maximize the utility function or to minimize the cost function. It should be noted that the utility function is concave, while the cost function is convex. Moreover, the problem involving the maximization of a concave function f can be similarly reformulated when the objective is to minimize function $-f$, which is convex (Deng *et al.* 2015).

Wang *et al.* (2016) use convex optimization to model to model the reduction of cloud data center energy costs. The approach they propose enables a cloud operator to jointly optimize the energy supply and the allocation of user workload in distributed data centers. Chen *et al.* (2014) consider a data center featuring energy storage devices. Their works aim to minimize the linear combination between the electricity cost and the average response time of user queries. Solutions based on convex optimization techniques are proposed to solve this problem.

7.4.3. *Markov decision process*

Markov decision process refers to sequential decision making based on the periodic or continuous observation of Markov dynamic systems. The actions of the latter are random, but the state transition probabilities have Markov properties.

Rao *et al.* (2010) model the uncertainty of the energy price and the workload processed by cloud data centers in a Markov decision process. They aim to reduce the electricity cost while providing users with quality of service. Yang *et al.* (2017) use a Markov decision process to model the resource supply of data center servers. They propose a self-management algorithm based on machine learning. Their proposal autonomously and dynamically provides resources to servers in order to reduce the electricity cost of data centers.

7.4.4. *Fuzzy logic*

Fuzzy logic is a subset of conventional Boolean logic, which was extended to deal with the concept of partial truth: truth values that range between "completely true" and "completely false". Fuzzy logic (Zadeh 1965) is a tool that deals with uncertain, imprecise or qualitative information in systems that are not defined with formal mathematical models. In theory, a fuzzy set indicates the progressive membership of an element to a set.

Let A be a fuzzy set, defined on a universe X by a membership function $\mu_A: X \rightarrow [0,1]$. μ_A indicates the membership degree of x to A. $\mu_A = 1$ indicates complete membership, $\mu_A = 0$ absolute non-membership and $0 < \mu_A < 1$ partial membership (the closer μ_A gets to 1, the stronger the membership of x to A).

Fuzzy logic covers a broad variety of functioning conditions. Moreover, fuzzy logic-based algorithms are able to deal with uncertainties and are robust in strong mutation environments (Zadeh 1965). Moreover, the inference process is simple, compared to other systems (Marrouchi and Ben Saber 2014), which leads to energy savings. In our case, this functionality is very interesting, as we deal with a real-time system in which user queries must be executed within short times while energy consumption is minimized.

Chopra *et al.* (2017) implemented an electricity monitoring and control system using fuzzy logic and cloud computing. This system is responsible for automated adjustment of the operating time of electrical devices. The system has humidity and temperature as input data and calculates the operating time of devices as output data. According to the results, the proposed system enables energy saving and supports energy efficient use. Benblidia *et al.* (2019) deal with the necessity of having an efficient method for task distribution in a fog–cloud architecture. Their proposal aims to send user queries to the best fog node, while meeting user preferences and the constraints of fog nodes. For this purpose, they used linguistic quantifiers and fuzzy quantified proposals efficiently grouping user preferences and fog constraints in order to classify the fog nodes from the most satisfactory to the least satisfactory. Simulation results indicated that the system distributes the tasks while meeting user preferences. It also offers a compromise between average user satisfaction, execution time and electric consumption.

7.5. Conclusion

The reduction of energy use and consumption costs and the reduction of carbon emissions in data centers are of high interest for both researchers and industrial actors. Smart grid development raised these efforts to a different level, integrating concepts that aim to improve the energy efficiency of data centers. DR, dynamic pricing and integration of DER are the most adopted approaches.

This chapter deals with the energy efficiency of data centers in a smart grid-cloud system. We first introduced the smart grid, its architecture integrating information and communication technologies, as well as its main functionalities. After having detailed the interaction of smart grid with cloud data centers, we summarized the main energy efficiency techniques used. Moreover, a comparative study of works on this subject was presented. This was followed by an overview on decision aiding techniques in a smart grid-cloud system.

Related to the studied work, it is worth noting that prices and the use of renewable energies are two main factors of energy efficiency in data centers. Two main themes are currently being explored for a more energy-efficient smart grid–cloud interaction. The first aims to run data centers in micro-grids to ensure reliability of electrical systems, while the second aims to distribute the workload among data centers in order to minimize the emission of greenhouse gases.

7.6. References

Allcott, H. (2009). *Real Time Pricing and Electricity Markets*. Harvard University Press, Cambridge.

Asmus, P. (2017). Data centers and advanced microgrids. White paper, NAVIGANT Research.

Avgerinou, M., Bertoldi, P., and Castellazzi, L. (2017). Trends in data center energy consumption under the European code of conduct for data centre energy efficiency. European Commission.

Benblidia, M.A., Brik, B., Esseghir, M., and Merghem-Boulahia, L. (2018). A game based power allocation in cloud computing data centers. *14th International Conference on Wireless and Mobile Computing, Networking and Communications*. WiMob, Limassol, 1–7.

Benblidia, M.A., Brik, B., Esseghir, M., and Merghem-Boulahia, L. (2019). Ranking fog nodes for tasks scheduling in fog-cloud environments: A fuzzy logic approach. *14th International Conference on Wireless Communications and Mobile Computing*. IWCMC, Tangier, 1–7.

Bera, S., Misra, S., and Rodrigues, J.J.P.C. (2015). Cloud computing applications for smart grid: A survey. *IEEE Transactions on Parallel and Distributed Systems*, 26(5), 1477–1494.

Bergaentzlé, C. and Clastres, C. (2013). Tarifications dynamiques et efficacité énergétique: l'apport des Smart Grids. *Presses de l'ISMEA*, XLVII(2), 348–363.

Cavdar, D. and Alagoz, F. (2012). A survey of research on greening data centers. *IEEE GLOBECOM*. 3–7 December 2012, Anaheim, California, USA, 3237–3242.

Chen, S., Wang, Y., and Pedram, M. (2014). Resource allocation optimization in a data center with energy storage devices. *IECON 2014 – 40th Annual Conference of the IEEE Industrial Electronics Society*. IECON, Dallas, 2604–2610.

Chiang, M., Ha, S., Chih-Lin, I., Risso, F., and Zhang, T. (2017). Clarifying fog computing and networking: 10 Questions and answers. *IEEE Communications Magazine*, 55(4), 18–20.

Chiang, M. and Zhang, T. (2016). Fog and IoT: An overview of research opportunities. *IEEE Internet of Things Journal*, 3(6), 854–864.

Chopra, P. and Bedi, R.P.S. (2017). Application of fuzzy logic in cloud computing: A review. *International Journal of Scientific Research Engineering & Technology (IJSRET)*, 6(11), 1083–1086.

Dai, X., Wang, J.M., and Bensaou, B. (2016). Energy-efficient virtual machines scheduling in multi-tenant data centers. *IEEE Transactions on Cloud Computing*, 4(2), 210–221.

Deng, R., Yang, Z., Chow, M., and Chen, J. (2015). A survey on demand response in smart grids: Mathematical models and approaches. *IEEE Transactions on Industrial Informatics*, 11(3), 570–582.

Deng, W., Liu, F., Jin, H., Li, B., and Li, D. (2014). Harnessing renewable energy in cloud datacenters: Opportunities and challenges. *IEEE Network Magazine*, 28(1), 48–55.

Ding, Z., Xie, L., Lu, Y., Wang, P., and Xia, S. (2018). Emission-aware stochastic resource planning scheme for data center microgrid considering batch workload scheduling and risk management. *IEEE Transactions on Industry Applications*, 54(6), 5599–5608.

Duong-Ba, T.H., Nguyen, T., Bose, B., and Tran, T.T. (2018). A dynamic virtual machine placement and migration scheme for data centers. *IEEE Transactions on Services*, 1–14.

Elmroth, E., Leitner, P., Schulte, S., and Venugopal, S. (2017). Connecting fog and cloud computing. *IEEE Cloud Computing*, 4(2), 22–25.

Endo, H., Kodama, H., Fukuda, H., Sugimoto, T., Horie, T., and Kondo, M. (2013). Effect of climatic conditions on energy consumption in direct fresh-air container data centers. *International Green Computing Conference Proceedings*, 1–10.

Gabr, A.Z., Helal, A.A., and Abbasy, N.H. (2018). Dynamic pricing; different schemes, related research survey and evaluation. *9th International Renewable Energy Congress.* Hammamet, 1–7.

Gandhi, A., Harchol-Balter, M., Das, R., and Lefurgy, C. (2009). Optimal power allocation in server farms. *SIGMETRICS Performance Evaluation Review*, 37(1), 157–168.

Gu, C., Hu, K., Li, Z., Yuan, Q., Huang, H., and Jia, X. (2016). Lowering down the cost for green cloud data centers by using ESDs and energy trading. *IEEE Trustcom/BigDataSE/ ISPA 2016*. August 23–26, 2016. Tianjin, China, 1508–1515.

Gu, C., Fan, L., Wu, W., Huang, H., and Jia, X. (2018). Greening cloud data centers in an economical way by energy trading with power grid. *Future Generation Computer Systems*, 78, 89–101.

Guerassimof, G. (2017). *Microgrids: pourquoi, pour qui?* Presses des Mines, Paris.

Guerassimof, G. and Maizi, N. (2013). *Smart Grids: au-dela du concept, comment rendre les réseaux plus intelligents.* Presses des Mines, Paris.

Gungor, V.C., Sahin, D., Kocak, T., Ergut, S., Buccella, C., Cecati, C., and Hancke, G.P. (2011). Smart grid technologies: Communication technologies and standards. *IEEE Transactions on Industrial Informatics*, 7(4), 529–539.

Hossain, E., Han, Z., and Poor, H.V. (2012). *Communication Architectures and Models for Smart Grid.* Cambridge University Press, Cambridge.

Joshi, Y. and Kumar, P. (2012). *Energy Efficient Thermal Management of Data Centers.* Springer-Verlag, New York.

Khan, A.A., Rehmani, M.H., and Reisslein, M. (2016). Cognitive radio for smart grids: Survey of architectures, spectrum sensing mechanisms and networking protocols. *IEEE Communications Surveys & Tutorials*, 18(1), 860–898.

Kiani, A. and Ansari, N. (2018). Profit maximization for geographically dispersed green data centers. *IEEE Transactions on Smart Grid*, 9(2), 703–711.

Kliazovich, D., Bouvry, P., Audzevich, Y., and Khan, S.U. (2010). Greencloud: A packetlevel simulator of energy-aware cloud computing data centers. *IEEE Global Telecommunications Conference GLOBECOM 2010.* 6–10 December 2010, Miami, Florida, USA, 1–5.

Liu, L., Sun, H., Li, C., Hu, Y., Li, T., and Zheng, N. (2018). Exploring customizable heterogeneous power distribution and management for datacenter. *IEEE Transactions on Parallel and Distributed Systems*, 29(12), 2798–2813.

Markovic, D.S., Zivkovic, D., Branovic, I., Popovic, R., and Cvetkovic, D. (2013). Smart power grid and cloud computing. *Renewable and Sustainable Energy Reviews*, 24, 566–577.

Marrouchi, S. and Ben Saber, S. (2014). A comparative study of fuzzy logic, genetic algorithm, and gradient-genetic algorithm optimization methods for solving the unit commitment problem. *Mathematical Problems in Engineering*, 2014, 1–14.

Mohsenian-Rad, A., Wong, V.W., Jatskevich, J., Schober, R., and Leon-Garcia, A. (2010). Autonomous demand-side management based on game-theoretic energy consumption scheduling for the future smart grid. *IEEE Trans. Smart Grid*, 1(3), 320–331.

Rao, L., Liu, X., Xie, L., and Liu, W. (2010). Minimizing electricity cost: Optimization of distributed Internet data centers in a multi-electricity-market environment. *Proceedings of the INFOCOM.* 14–19 March 2010, San Diego, California, USA, 1–9.

Samadi, P., Mohsenian-Rad, H., Schober, R., and Wong, V.W. (2012). Advanced demand side management for the future smart grid using mechanism design. *IEEE Trans. Smart Grid*, 3(3), 1170–1180.

Sharma, N.K. and Reddy, G.R.M. (2019). Multi-objective energy efficient virtual machines allocation at the cloud data center. *IEEE Transactions on Services Computing*, 12(1), 158–171.

Somani, A. and Joshi, Y.K. (2009). Data center cooling optimization: Ambient intelligence based load management. *Proceedings of the ASME Summer Heat Transfer Conference 2009*. July 19–23 2009, San Francisco, USA.

Terzija, V., Valverde, G., Cai, D., Regulski, P., Madani, V., Fitch, J., Skok, S., Begovic, M.M., and Phadke, A. (2011). Wide-area monitoring, protection, and control of future electric power networks. *Proceedings of the IEEE*, 99(1), 80–93.

Vardakas, J.S., Zorba, N., and Verikoukis, C.V. (2015). A survey on demand response programs in smart grids: Pricing methods and optimization algorithms. *IEEE Communications Surveys & Tutorials*, 17(1), 152–178.

Wang, D., Ren, C., and Sivasubramaniam, A. (2013). Virtualizing power distribution in datacenters. *SIGARCH Computer Architecture News*, 41(3), 595–606.

Wang, H., Huang, J., Lin, X., and Mohsenian-Rad, H. (2016). Proactive demand response for data centers: A win-win solution. *IEEE Transactions on Smart Grid*, 7(3), 1584–1596.

Wang, H. and Ye, Z. (2016). Renewable energy-aware demand response for distributed data centers in smart grid. *IEEE Green Energy and Systems Conference (IGSEC)*. IEEE, Long Beach, 1–8.

Wang, Y., Lin, X., and Pedram, M. (2014a). Coordination of the smart grid and distributed data centers: A nested game-based optimization framework. *ISGT 2014*. IEEE, Washington, 1–5.

Wang, Y., Saad, W., Han, Z., Poor, H.V., and Baar, T. (2014b). A game-theoretic approach to energy trading in the smart grid. *IEEE Transactions on Smart Grid*, 5(3), 1439–1450.

Yan, Y., Qian, Y., Sharif, H., and Tipper, D. (2013). A survey on smart grid communication infrastructures: Motivations, requirements and challenges. *IEEE Communications Surveys & Tutorials*, 15(1), 5–20.

Yang, J., Zhang, S., Wu, X., Ran, Y., and Xi, H. (2017). Online learning-based server provisioning for electricity cost reduction in data center. *IEEE Transactions on Control Systems Technology*, 25(3), 1044–1051.

Yildizoglu, M. (2011). *Introduction à la théorie des jeux*. Dunod, Paris.

Yu, L., Jiang, T., and Cao, Y. (2015). Energy cost minimization for distributed internet data centers in smart microgrids considering power outages. *IEEE Transactions on Parallel and Distributed Systems*, 26(1), 120–130.

Yu, L., Jiang, T., Cao, Y., and Qi, Q. (2014). Carbon-aware energy cost minimization for distributed Internet data centers in smart microgrids. *IEEE Internet of Things Journal*, 1(3), 255–264.

Yu, L., Jiang, T., and Zou, Y. (2016). Real-time energy management for cloud data centers in smart microgrids. *IEEE Access*, 4, 941–950.

Zadeh, L. (1965). Fuzzy sets. *Information and Control*, 8(3), 338–353.

8

Toward New Intelligent Architectures for the Internet of Vehicles

Léo MENDIBOURE[1], Mohamed Aymen CHALOUF[2] and Francine KRIEF[3]

[1] *LaBRI, Bordeaux, France*

[2] *IRISA, Rennes, France*

[3] *ENSEIRB-MATMECA, Bordeaux, France*

8.1. Introduction

The reference architecture of Cooperative-Intelligent Transport Systems (C-ITS) (ETSI 2010) was defined by the joint work of various standardization organizations: IEEE, ISO, ETSI, etc. Structured along three main axes (management, control, security), this architecture must enable the deployment of a large-scale vehicular communication system. As proven by Kaiwartya *et al.* (2016), in the current state it has various limitations: capacity to process a significant data volume, interoperability between various networks, guaranteed quality of service (QoS), guaranteed security and respect of private life, etc. This is why vehicular networks evolved toward a new paradigm: the Internet of Vehicles (IoV).

Relying on the principles of the Internet of Things (IoT), IoV must enable addressing the limitations of Vehicular Ad hoc NETworks, also known as "VANET". For this purpose, IoV relies on the development of new types of communication: vehicle to pedestrian (V2P), vehicle to object (V2O), etc. This must make possible the improvement of road safety services, the development of global

Intelligent Network Management and Control,

traffic management services as well as the design of new entertainment services (multimedia, advertising, etc.).

The communication architecture that should enable the ascent of the IoV has not yet been normalized. This architecture must meet many criteria: safety and protection of private life, simplified management, optimized dissemination of information, scaling, etc. (Kaiwartya *et al.* 2016). Various architectures aiming to make possible these various improvements, based on the reference C-ITS architecture, were proposed in the literature. These architectures rely on the integration of various technologies: network function virtualization (NFV), software-defined network (SDN), Blockchain, edge computing (EC), etc.

All these architectures share a common point: the idea of integrating artificial intelligence (AI) techniques in the control plane. Indeed, whatever the chosen approach (NFV, SDN, EC, etc.), knowledge sharing and intelligent decision making are presented as essential points. The use of AI, in addition to other technologies, could bring many improvements to the control plane: dynamic routing, prediction of vehicle movement and congestion control.

However, the set of architectures described in the literature focuses only on the improvement of the control plane. But intelligent decision making could also reinforce security plane and private life protection (intrusion detection, secure routing, etc.) as well as the architecture management plane (load balancing, allocation of virtual resources, etc.). Moreover, as suggested in Mendiboure *et al.* (2019a), the addition of a fourth plane, the knowledge plane (KP), in the service of all the other planes (management, control and security) could have many advantages.

The objective of this chapter is to describe and compare the work aiming to improve the operation of vehicular networks by AI integration in the control plane. Then we define an architecture enabling this work and offering intelligent services, not only to the control plane, but also to the management and security planes. Using existing work, we present for each of these planes the advantages of using AI techniques and a list of challenges to be addressed is provided.

First of all, section 8.2 presents the context: the principle of the IoV and also its applications. Then section 8.3 presents the various works integrating AI techniques in the vehicular communication architecture. This same section explains the interest of the definition of a new architecture for IoV. As a continuation, section 8.4 describes in detail the proposed architecture, its main components, as well as its advantages. Finally, section 8.5 presents the stakes of the proposed architecture and section 8.6 presents the conclusion and the perspectives concerning the integration of AI techniques in the vehicular communication architecture.

8.2. Internet of Vehicles

This section presents the characteristics of the IoV as well as the applications made possible by this new vehicular architecture.

8.2.1. *Positioning*

As indicated in Kaiwartya *et al.* (2016), the development of the IoV has three main objectives:

– address the limitations of vehicular *ad hoc* networks, namely:

- lack of interoperability: the current vehicular architecture does not enable the interconnection of heterogeneous networks (Wi-Fi, cellular, Li-Fi, etc.), thus preventing the deployment of global and reliable intelligent transportation services;

- absence of Internet connectivity: the current vehicular architecture does not guarantee Internet connectivity to vehicles. Consequently, the development of commercial services (advertising, entertainment, etc.) or of efficient traffic management services is impossible;

- limited intelligent decision making: due to limited Internet connectivity and reduced computing/storage capacities of vehicles, the processing of significant data volumes is impossible;

- lack of exchanges: vehicular *ad hoc* networks being designed to enable vehicle interconnection, the integration of new types of connected objects (cameras, telephones, etc.) is not an option, although it could have many benefits (detection of pedestrians and cyclists, creation of cooperative map, etc.);

– address the increase in the number of traffic accidents: intelligent transport systems aim first of all to provide and improve road safety. Since the number of traffic accidents and fatalities is constantly increasing (Kaiwartya *et al.* 2016), the development of road safety services is essential and must rely on an architecture enabling the development of global services;

– provide new opportunities: integration of new objects in the vehicular networks, interoperability between various networks and guaranteed Internet connectivity could enable the generation of significant data volumes, the provision of new services and consequently attracting new investors: service providers, car manufacturers, etc.

8.2.2. *Characteristics*

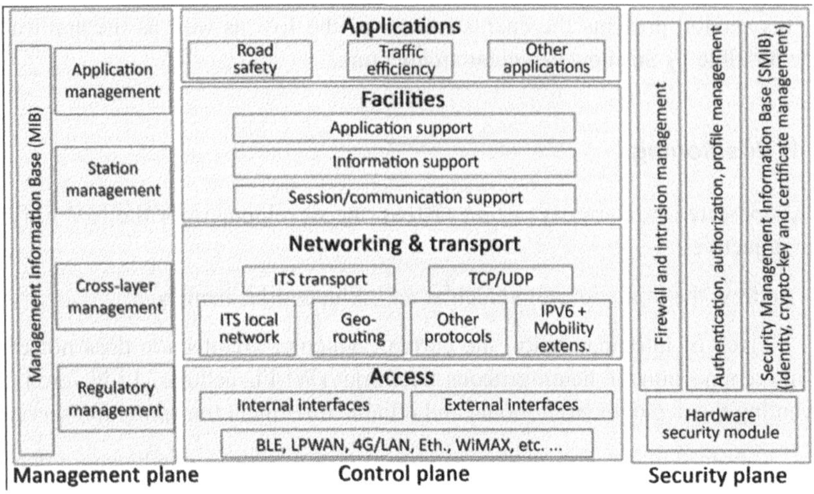

Figure 8.1. *Presentation of C-ITS architecture*

The development of the IoV and the improvement of vehicular communication network performances rely on two important concepts:

– integration of new types of objects in the vehicular networks through new types of communication, which is referred to as vehicle to everything (V2X):

- vehicle to pedestrian (V2P);

- vehicle to object (V2O);

- vehicle to grid (V2G);

- vehicle to cloud (V2C);

– integration of new technologies in the C-ITS architecture (Figure 8.2), each of them making it possible to overcome certain limitations of vehicular *ad hoc* networks:

- NFV: provides virtualization of network functions and services leading to cost reduction (standard material), more rapid deployment and simplified scalability (flexibility);

- SDN: a technique for uncoupling the control plane and the data plane, enabling network programmability and an overall view, and high adaptation to the context (dynamicity);

- EC: based on the deployment of computing and storage capacities on the edge of the network, extension of cloud capacities, which should limit the latency times (better user experience) and unload the network core (scalability);

- blockchain: a distributed technology enabling transparent, secure and unfalsifiable information storage and exchange, which facilitates the development of efficient, robust and scalable security solutions;

- AI: involves techniques that should enable the improvement of the existing architecture, and also newly integrated technologies (EC, NFV, SDN) through decision-aiding and process automation methods.

8.2.3. *Main applications*

Addressing the limits of vehicular *ad hoc* networks, integrating new types of objects and enabling Internet connectivity, the IoV facilitates many new applications, which can be classified in various categories (Mendiboure *et al.* 2019b):

– transport management: real-time traffic management, parking place reservation, electric vehicle charging points, etc.;

– road safety: intersection management, cooperative avoidance of collisions, detection of obstacles, etc.;

– vehicle management: personal assistant, remote maintenance system, etc.;

– driver assistance: advanced driver assistance systems (ADAS), parking assistance, creation of cooperative maps, etc.;

– connected health: tiredness detection, user comfort, medical assistance, etc.;

– new transportation modes: platooning, creation of connected vehicle convoys or car sharing, new modes of transportation aimed at reducing the number of vehicles on the roads and fuel consumption, and therefore pollution.

8.3. IoV architectures proposed in the literature

To overcome the limits of vehicular *ad hoc* networks and offer new services, many evolutions of C-ITS architecture were proposed in the literature. This section aims to present these various architectures by highlighting an essential point: the integration of AI techniques in these systems. Therefore, the advantages and limits of these architectures are evaluated based on their automation level.

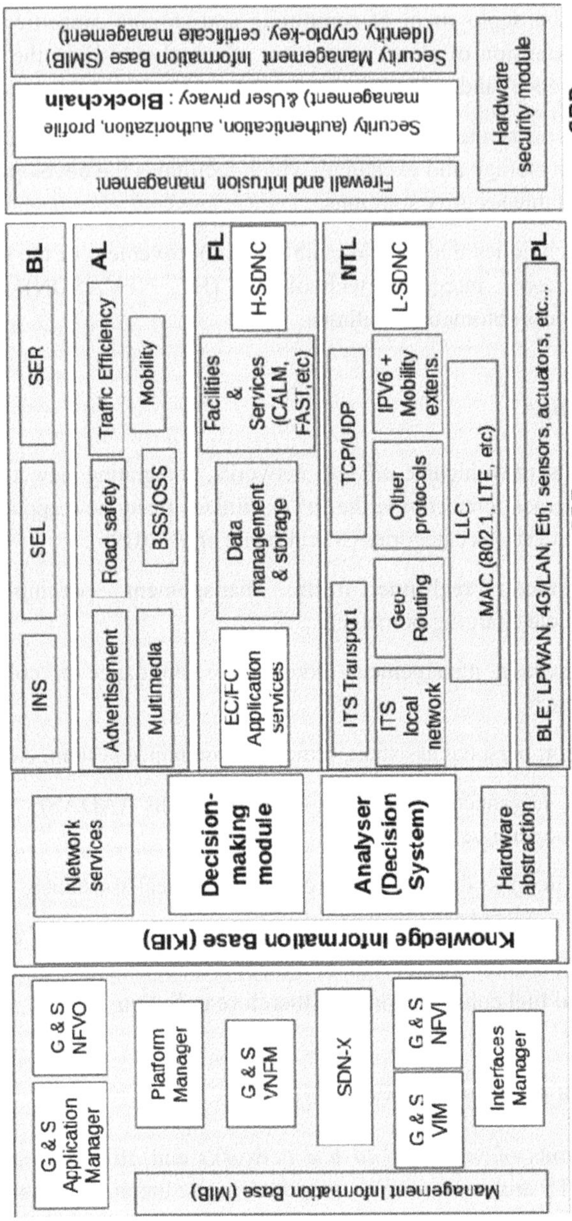

MOP: Management and Operation Plane; KP: Knowledge Plane; CDP: Control and Data Plane; SPP: Security and Privacy Plane
BL: Business Layer; AL: Application Layer; FL: Facilities Layer; NTL: Network and Transport Layer; PL: Perception Layer

Figure 8.2. *Presentation of IoV architectures described in the literature*

8.3.1. *Integration of AI techniques in a layer of the control plane*

The first approach proposed in various papers involves the integration in the vehicular architecture of a new layer, namely an AI layer enabling the recovery, storage, analysis and processing of data from lower layers (Kaiwartya *et al.* 2016; Aliyu *et al.* 2017; Borcoci *et al.* 2017).

Aliyu *et al.* (2017) (Figure 8.2(2)) define a new architecture composed of four layers and specifically aimed at the development of cloud computing services in a vehicular environment. The AI layer should enable the processing of data from lower layers (perception, coordination) and thus render possible applications requiring the processing of massive volumes of data, notably for real-time traffic management. This layer also aims to improve the operation of cloud services by increasing the computing capacity of the vehicles using cloud (VuC) services and by optimizing the use of the vehicular cloud computing (VCC) resources.

A similar architecture, intended to serve as a basis for the IoV, is proposed by Kaiwartya *et al.* (2016) (Figure 8.2(6)). Using various AI techniques (expert systems, VCC, massive data analysis), this layer has the same objectives as that presented by Aliyu *et al.* (2017): processing of massive volumes of data and application operation improvement. Nevertheless, according to the authors, this layer could also enable the commercialization of vehicular networks. Indeed, the definition and development of economic models (*Business* layer) for service providers could rely on the statistical analysis of application use data provided by the AI layer.

This definition of a five-layer architecture is resumed by Borcoci *et al.* (2017) (Figure 8.2(3)). Nevertheless, it is extended in two directions: integration of EC capacities and integration of SDN in the architecture. EC capacities may enable the deployment of intelligent services as close as possible to the user, thus improving the user experience quality and limiting the transmission of information. Moreover, the use of SDN may lead to improving the management of these resources: positioning of applications and reservation of resources (bandwidth, storage and computing capacity, etc.).

8.3.2. *Integration of AI techniques in several layers of the control plane*

However, the interest of AI is not limited to cloud resources management and processing of information for applications.

Contreras-Castillo *et al.* (2017) (Figure 8.2(4)) define a seven-layer architecture for IoV. Intelligence is integrated here at various levels: the preprocessing layer, communication layer and control layer. At the level of the preprocessing layer, data processing should make it possible to limit the amount of data transmitted and improve relevance, which would consequently reduce network load. At the level of the communication layer, AI must enable an optimal choice of the transmission technology depending on the required QoS. Finally, at the level of the control layer, AI must enable a management of service providers and data exchange, as it was already proposed in previous works (section 8.3.1).

This idea of applying AI techniques at various levels can also be found in Chen *et al.* (2018) (Figure 8.2(5)). The authors introduce the idea of cognitive Internet of Vehicles (CIoV). According to them, the absence of intelligent decision making would limit the influence of IoV: suboptimal resource management, lack or reliability of the decision process, weak marketing potential, etc. Hence, AI is applied here not only to network management (network traffic management, safety management, resource allocation), but also to a protection of road users: vehicle self repair, driver health monitoring, driver safety control, emotions analysis, etc. AI application is consequently extended to various levels (communication, cognition and control) and also to various domains (human and network).

8.3.3. *Definition of a KP associated with the control plane*

To allow for interlayer decision making, Jiacheng *et al.* (2016) (Figure 8.2(1)) propose the addition of a KP. In this SDN-based architecture, the KP must enable the processing of information coming from the lowest layers (data layers) depending on the needs of other layers (application, control). Hence, this plane must be of help in addressing the main challenges related to SDN integration: management of resource mobility and heterogeneity through various functions (packet transmission control, network virtualization, handover management, etc.).

8.3.4. *Comparison of architectures and positioning*

The work presented in this section aims to improve the vehicular communication architecture through the integration of AI techniques.

The set of works presented aims to enable the improvement in control plane performances (Table 8.1). Some of these works focus on the evolution of only one layer, the applicative layer, through the addition of one AI layer for data preprocessing and analysis serving applications (Kaiwartya *et al.* 2016; Aliyu *et al.* 2017; Borcoci *et al.* 2017). Others propose multilayer approaches intended also to a better transportation of data (interoperability, routing, bandwidth optimization)

(Contreras-Castillo *et al.* 2017; Chen *et al.* 2018). Only one of these papers (Jiacheng *et al.* 2016) proposes the addition of a KP that may enable interlayer decision making and centralized and optimized information processing.

Architecture	Contribution			Multilayer	Knowledge plane
	Management	Control	Security		
Aliyu *et al.* (2017)	Cloud	Yes	No	No	No
Kaiwartya *et al.* (2016)	Cloud	Yes	No	No	No
Borcoci *et al.* (2017)	Cloud	Yes	No	No	No
Contreras-Castillo *et al.* (2017)	Non	Yes	No	Yes	No
Chen *et al.* (2018)	Yes	Yes	No	Yes	No
Jiacheng *et al.* (2016)	No	Yes	No	Yes	Yes
Proposal	Yes	Yes	Yes	Yes	Yes

Table 8.1. *Comparison of various proposed architectures*

Some works also propose the use of AI to support the management plane (cloud resources management, resource management on the edge of the network, positioning of applications) (Kaiwartya *et al.* 2016; Aliyu *et al.* 2017; Borcoci *et al.* 2017; Chen *et al.* 2018). Nevertheless, Jiacheng *et al.* (2016) do not address this question and focus on the interactions between the SDN control plane and the KP. Moreover, the idea of interlayer decision making and optimization at several levels has until now been proposed only in support of the control plane. Furthermore, none of the previously mentioned works addresses the idea of applying AI to support the security plane and protect private life (intrusions, routing, etc.) irrespective of its reliability.

To address these limitations, the following part of this document defines a new architecture for the IoV and studies the main advantages, as well as the main barriers, in terms of control, management and security and private life.

8.4. Our proposal of intelligent IoV architecture

This section presents the proposed IoV architecture as well as the various plans it includes (control, management, security and private life, and knowledge). Moreover, it analyzes the various AI applications for control, management and security and private life presented in the literature.

8.4.1. *Presentation*

The proposed intelligent IoV architecture is presented in Figure 8.3. It includes the following four planes:

– management and orchestration plane (MOP): this plane is in charge with the global management of architecture and resources;

– security and privacy plane (SPP): this plane offers security and private life protection services to all the other planes: authorizations, access control, profile management, etc.;

– KP: this plane offers information processing and analysis and decision making services to all the other planes;

– control and data plane (CDP): this plane is in charge of recovery, transportation and optimal distribution of information.

This architecture is the first to propose a KP for the IoV. The first work on the KP dates from 2003 and aimed to enable networks to make decisions, particularly management decisions, without human intervention (Clark *et al.* 2003). The management tasks were accomplished by the network itself, each autonomous entity being able to self-configure, self-optimize, self-protect and self-repair (Krief 2010). Due to implementation difficulties, the KP realizations were limited to concept proofs, but given the current advances, mainly in the field of AI, its deployment can be expected today.

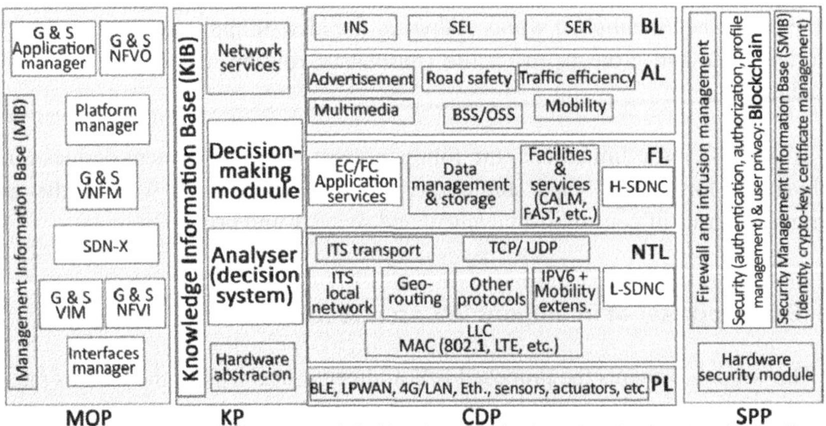

MOP: Management and Operation Plane; KP: Knowledge Plane; CDP: Control and Data Plane; SPP: Security and Privacy Plane; BL: Business Layer; AL: Application Layer; FL: Facilities Layer; NTL: Network and Transport Layer; PL: Perception Layer

Figure 8.3. *Our proposal*

8.4.2. *A KP for data transportation*

Vehicle mobility is very important for data transportation. Indeed, the variation in the number of vehicles with time in a given space involves constant optimization of data transportation to enable high performances: routing, congestion, etc. The application of AI techniques to the control plane of architecture is proposed for various applications, particularly vehicle flow motion prediction, congestion control and dynamic routing.

Vehicle flow motion prediction is a first important point. Indeed, it must enable the improvement of many ITS applications: management of road traffic congestion, reduction of greenhouse gas emissions, reduction of fuel consumption or improvement of services based on localization. Huang *et al.* (2014) were the first ones to propose a deep learning-based approach for vehicle flow motion prediction. This solution, based on a stacked restricted Boltzmann machines, integrating the idea of multitask learning, aims to improve the performances of methods based on previously used static networks and to prove the interest of a dynamic approach. Nevertheless, the proposed system could be improved. Indeed, traffic prediction is not coupled here with time information. Similarly, Lv *et al.* (2015) proposed another deep learning-based approach that should enable fine prediction of vehicle motion. Because of unsupervised learning of an auto-encoder model, this approach should also enable nonlinear spatial and temporal correlations by analyzing the data flow. The comparison between this method and other existing approaches (back propagation, ransom walk, support vector machines) proves its efficiency in terms of road traffic prediction. Finally, in Ide *et al.* (2015), the authors consider both the vehicle flow motion prediction problem and that of Long-Term Evolution (LTE) communications connectivity. Hence, because of the physical data provided by the vehicles and the use of Poisson dependency networks, they build a model enabling them to predict vehicle motion. Owing to information on cellular connectivity measurement, the proposed system also enables, using Poisson regression trees, the distributed prediction of cellular connectivity for each vehicle. Thanks to this, it should be possible to improve the vertical handover process and optimize machine-to-machine communications.

A second important point is congestion control. Indeed, to limit the packet loss and the end-to-end transmission delay and to increase the reliability of vehicular networks, limiting data congestion is essential. Congestion control strategies can rely on various approaches aimed at optimizing various parameters: power management, throughput management, prioritization and scheduling, etc. The idea of intelligent congestion control, to improve these solutions, was proposed in many articles. For example, Taherkhani and Pierre (2016) discuss a solution based on a k-means partitioning that should enable congestion management in intersections using a roadside unit (RSU) to control the congestion level of radio channels. These RSUs recover the data transmitted by the set of vehicles at the intersection level and group them depending on various parameters: size and type of messages, validity, position of the receiver, distance between the vehicle and the road side infrastructure, etc. Consequently, depending on the available resources, communication parameters are assigned to each of these groups: transmission delay, transmission power, congestion window size, etc. Consequently, this enables congestion limitation. Nevertheless, the solution proposed has various limitations, notably the computing capacity and the latency related to this information processing. For the proper operation of cooperative road safety services, the vehicles send short messages to surrounding vehicles (beacons) containing varied information: position, speed, etc. In an area with a significant number of vehicles, the transmission of these regular messages may lead to congestion problems. This is why Toutouh and Alba (2016, 2018) proposed swarm intelligence–based congestion control methods aimed at optimizing the frequency of these messages. Through these methods the use rate of radio communication channels is maintained at a level enabling the proper operation of the network, congestion prevention, while maximizing the number of exchanged messages and, consequently, improving the reliability of the information received by each vehicle. However, as highlighted by these authors, the performances of these two algorithms remain still to be proven within a real environment.

Finally, communication between two vehicles requires a calculation of the path taking into account the characteristics of various links (life duration, bandwidth, delay, etc.). Many approaches based on AI techniques were proposed for making an optimal calculation. For example, Zhang *et al.* (2018a) proposed a solution based on genetic algorithms enabling an optimal routing aimed at ensuring high QoS. These algorithms try to rapidly find an approximate solution to an optimization problem. Using information related to vehicle traffic (speed, direction, etc.), the authors describe a mechanism for the dynamic computation of the optimal communication path between two vehicles. The proposed solution should enable the management of connectivity dynamicity (motion of vehicles, link break) and the optimization of the packet transmission delay. Nevertheless, the proposed process is complex, and the optimization of the time for searching the optimal path should be considered. In Lai *et al.* (2015), a machine learning-based approach is proposed. The authors consider three main problems of data dissemination: vehicle motion prediction, evaluation of the transmission capacity of the equipment and evaluation of the transmission direction. The study of these various problems is done at the RSU level, which share information concerning the position of vehicles and the capacity of their communication links. An intelligent computation of the optimal path is then done by the system. The authors want this way to prove the relevance of using AI techniques for routing, ensuring a packet loss ratio and a latency below those offered by the existing solutions. Finally, Yao *et al.* (2018) describe a mechanism that should enable the determination of future positions of the vehicles and finding the optimal path to reach them. For this purpose, the authors propose a system based on a hidden Markov model and on the history of past motions of the vehicles. Indeed, by analyzing the places previously visited by these vehicles and making a time–space correlation, it is possible to determine where the vehicles are heading to and, consequently, what paths they use. Using this approach, it seems possible to improve the packet transmission delay and to reduce losses. However, to verify the performances of the solution proposed by the authors, implementation in the real environment is required.

8.4.3. *A KP for IoV architecture management*

Vehicular communication architecture must enable optimal resource management: computing, storage and communication. This management must be present at various levels to enable load balancing between various heterogeneous networks, between various operators as well as between various computing servers. This should enable high performances for all the users, offering the possibility to minimize delay and packet loss and to maximize bandwidth. Similarly to data transmission, for an optimization of resource use, an AI approach was proposed in many articles. The existing works can be classified into three large categories corresponding to various contributions: load balancing between various heterogeneous networks, load balancing for intervehicular communications and management of computing and storage resources (EC).

Due to the motion of vehicles, there are areas and, consequently, base stations that may have to deal with a large number of communications, driving a reduction of QoS. This is why intelligent solutions were developed. These solutions enable load balancing and vertical control in vehicular networks. For example, Li *et al.* (2017) proposed a solution aimed at enabling efficient association of the user with heterogeneous base stations due to reinforcement learning. Indeed, in order to guarantee good QoS, an optimized distribution of vehicles between various base stations is required. Hence, the association is first established depending on the current information of the base stations and their capacities. Then associations are optimized depending on the information collected by these base stations (peak traffic times, motion of vehicles, etc.), making it possible to determine the number of vehicles that should be associated in a given area and, consequently, how to associate them. To prove the benefits of this approach, the authors compared it to currently used methods: maximal signal to noise ratio (max-SINR) and distributed dual decomposition optimization. In the case of heterogeneous networks, load balancing can also be used to guarantee maximal QoS at any moment, as proven in Xu *et al.* (2014). The authors of this article also propose the use of a reinforcement learning method (fuzzy Q-learning) that should offer significant adaptability to the evolution of traffic conditions. Indeed, it must enable the determination of the optimal network for data transmission depending on four parameters: signal to noise ratio, number of vehicles connected to the concerned network, speed of the vehicle and amount of data to be transmitted. The authors prove the relevance of this service for multimedia applications. Because of the proposed load balancing technique, throughput is improved and low latency is ensured. Finally, in an attempt to limit the deployment of base stations and to best use the available resources, Zheng *et al.* (2016) defined a system expected to help optimize the packet transmission delay in a virtual environment (SDN). This system, relying on Markov chains and Bellman equations, is composed of two main parts. The first is a macroscopic management of virtual resources allocation (MaVRA) based on the analysis of global and time-extended parameters, especially traffic density. As for the second part, it involves a microscopic management of virtual resources allocation (MiVRA) depending on rapidly evolving parameters, notably the state of the queues of network equipment and the state of communication channels. Through a simulation, the authors of this article prove the gains in terms of delay of the proposed approach.

Certain applications, notably those related to road safety, have strong latency constraints and rely mainly on intervehicular communications. However, to maintain high QoS, it is important to use the set of available resources optimally. This is why Ashraf *et al.* (2016) introduced a system enabling load balancing between various vehicle-to-vehicle (V2V) paths. For this purpose, the authors define a mechanism introducing the formation of groups of vehicles depending on their movement (position, direction and speed). The formation of these groups enabled the optimization of intergroup exchanges. This optimization also occurs inside these groups because of the formation of pairs of vehicles depending on the stability and QoS of these communication links. Hence, this system enables the best use of the available V2V links. Similarly, Ye *et al.* (2019) define a decentralized mechanism enabling a vehicle to optimally use the available communication resources. This system, based on reinforcement learning and deep learning, meets the V2V communication constraints, while minimizing the interferences with vehicle-to-infrastructure (V2I) communications.

Finally, in order to offload the core of the network and offer a lower latency, as well as a higher throughput, many works advance the idea of deploying information storage and processing servers at network edge (EC). There are, nevertheless, many problems related to the use of these resources, such as whether these resources should be available at the level of the RSU (base stations) or inside the vehicles.

This is why Sun *et al.* (2018) are interested in the sharing the computing resources between vehicles. Hence, if one vehicle has resources that it does not use at a given moment, these resources could be used by the surrounding vehicles. The problem of the QoS of this type of service however remains, and this is why the authors propose a system that should enable the latency time minimization during the outsourcing of data processing. This approach relies on a task replication algorithm based on game theory (combinatorial multi-armed bandit theory). This system should enable each vehicle intending to outsource computing tasks to know how much time would take to each of the surrounding vehicles to complete this task. Thanks to this, it is possible for these vehicles to select a vehicle offering sufficient QoS for the considered application. The authors also consider the question of task replication, when the density of vehicles is significant, and the performance evaluation of this system. These tasks can also be dealt with at the RSU level. However, to avoid overloading an equipment, interequipment collaboration is essential. This is why Li *et al.* 2019 introduced reinforcement-based learning to enable load balancing between various roadside servers. This system relies on road traffic analysis and future road traffic prediction to estimate the load level of each server of roadside equipment and, consequently, to optimize task distribution (data processing, storage, etc.), particularly for HD (high-definition) applications that should improve road traffic conditions.

8.4.4. *A KP for securing IoV architecture*

In order to guarantee road user safety and prevent the actions of malicious entities, securing vehicular communications is also an essential issue. Indeed, sending erroneous messages, deleting messages or still disturbing communications could strongly deteriorate the performances of IoV system, and even render it dangerous for road safety. Because of the integration of AI techniques, the safety level of the IoV architecture could be improved, particularly through intrusion detection, prevention of routing attacks and protection against denial of service (DoS) attacks.

Intrusion detection should facilitate the identification of abnormal behaviors at various levels: vehicle, roadside equipment, etc. Kang and Kang (2016) propose a system that should enable the detection of intrusions inside the vehicle, using deep learning techniques. In particular, this system should enable the detection of abnormal behaviors inside the controller area network (CAN) data bus of the vehicle. This CAN bus is a system used to establish a communication between the various components of the vehicle. Its security is consequently essential for the operation of the autonomous vehicle. According to the evaluation conducted by the authors, the proposed unsupervised classification method and the definition of characteristic vectors based on the analysis of packets sent through the vehicle should enable efficient detection of real-time attacks. Zeng *et al.* (2018) aim to propose a system that uses machine learning to locally and globally detect intrusions. Global detection is done by RSUs, which have significant computing capacity and a global view of the network, due to their exchanges. Local detection is done at the vehicle level. Vehicles being organized in cluster, this system aims to identify abnormal behaviors of two types of vehicles: cluster heads (CHs) and vehicles exchanging of other clusters (multipoint relay [MPR]). Data concerning CH context and behavior are sent to RSUs and analyzed using a neural network in order to determine if the behavior of the CH is normal or malicious. Locally, vehicles are in charge of controlling the behavior of vehicles communicating with other clusters (MPR). Their computing capacities being limited, these vehicles compare the number of packets received and transmitted by these MPR vehicles. Hence, the combination of these two systems (local and global) must enable securing vehicular communications and deleting malicious vehicles. Finally Alheeti *et al.* (2015) compare the performances of various AI techniques for intrusion detection and abnormal behaviors of vehicles (gray holes) in vehicular networks. The first approach is based on a support vector machine (SVM) system and the second on a feedforward neural network (FFNN) system. This comparison proves the high efficiency of neural networks in anomaly detection and the limitation of the number of false negatives. However, the performances of this approach in terms of delays prove to be significantly lower than those of the SVM approach.

The detection and prevention of DoS attacks is another essential point. Indeed, the perturbation of vehicular communication services can hinder the exchanges between vehicles or between vehicles and infrastructure and may jeopardize the road users' safety. This is why Nyabuga *et al.* (2016) propose the detection of attacks using particle swarms optimization. Indeed, for the detection of DoS attacks on the V2I and V2V communications, the swarm approach may yield detection probabilities higher than those of the approaches using genetic algorithms, as proven by the authors. Alheeti *et al.* (2016) propose the first approach based on fuzzy Petri nets for the detection of intrusion and DoS attacks in a vehicular environment. The proposed approach must enable, because of the extraction of various characteristics (number of forwarded packets, number of discarded packets), the identification of external attacks (at the level of other vehicles) and internal attacks (inside the vehicle) with high probability and a limited false alarm rate. Nevertheless, it is still possible to improve the proposed solution (extension to other attacks, evaluation and use of resources). Finally, in Lyamin *et al.* (2018), the authors focus on the detection of radio jamming–based attacks in platoons. Indeed, the formation and proper operation of these platoons rely essentially on the exchange of information between vehicles and a perturbation of these communications, readily implemented, could pose a real danger. Hence, the authors first compare the existing reference methods based on models or data. Second, they propose a new hybrid method based on the statistical analysis of the network traffic and on data exploration methods. This new intelligent approach should make it possible to reach acceptable performances, even in cases of variable jitter during the emission of information messages. The same article undertakes an interesting discussion related to the application of deep learning to vehicular networks and the current limitations of these approaches, notably the amount of data required for learning.

Finally, during the computing of the communication paths, in order to avoid network disruption, it seems important to take into account the potentially malicious or infected nodes in order to eliminate them from the routing process. Hence, Krundyshev *et al.* (2018) focus on the malicious nodes transmitting false routing information and indicating that they provide the shortest communication path to the destination node, without this being the case. There are many routing attacks in vehicular networks: DoS, black hole, wormhole, Sybil, etc. The proposed solution should make it possible to prevent these various attacks by analyzing node behavior. For this, it relies on a confidence model and on a swarm intelligence algorithm (intelligent water drops algorithm). Because of a definition of a confidence index for each vehicle, based on the opinion of other vehicles, and because of a rapid transmission of this information, the proposed solution makes it possible to maintain a low transmission delay and high throughput and received packet percentage for various types of attacks (wormholes and black holes). Similarly, Zhang *et al.* (2018b) try to define an approach leading to high packet delivery rates and low throughputs in attack situations, without generating significant additional communication costs. This method relies on an adaptive routing algorithm AntHocNet, to which a fuzzy logic detection system was applied in order to exclude a decision and routing process, as the nodes may prove to be malicious. To identify the malicious nodes, each vehicle analyzes the behavior of surrounding vehicles depending on various parameters, notably the transmitted (received/sent) packet ratio over a certain period of time. This is why this method makes it possible to improve the performances of AntHocNet algorithm in case of black hole attacks. However, the scalability of this solution and the analysis of its performances for other types of attacks remain to be studied.

8.5. Stakes

The architecture that was proposed here, relying on the definition of a KP that serves all the other planes (control, management, security and private life) is expected to have many benefits: intrusion detection, load balancing, congestion control, vehicle motion prediction, etc. Nevertheless, the implementation of the proposed architecture depends on overcoming many obstacles, notably the security and the protection of private life, the complexity of computing methods, distributed learning and vehicle flow motion.

8.5.1. *Security and private life*

As indicated in Liang *et al.* (2019), although intelligent approaches may have many benefits for the security plane (intrusion detection, secure routing, etc.), the integration of AI techniques in vehicular architecture may raise serious security and confidence questions. Indeed, since training these algorithms relies on the analysis of existing data as well as on the integration of erroneous data or data manipulated by a malicious entity, these may modify the expected behavior of the intelligent system. Hence, in an environment as sensible as the vehicular environment, the study of reliability and robustness of these AI techniques seems essential. Moreover, the development of these approaches requires sharing user and processing power outsourced on the edge of the network or in the cloud. The development of solutions ensuring protection of user private life (anonymity, processing as close as possible, etc.) is also an essential question.

8.5.2. *Swarm learning*

Swarm learning techniques operate based on the collection of information by each node of the network and by information exchange between these nodes. Indeed, cooperation is essential so that each node has a global or partially global view. But in a vehicular environment, information transmission serving this intelligence may require significant communication resources that may not be available. Indeed, swarm learning approaches intended for vehicular communications such as Toutouh and Alba (2016, 2018) or Nyabuga *et al.* (2016) do not take into account the limitations related to vehicular environment. Moreover, applying these methods to vehicular environment requires considering the idea of cooperation systems that take into account the physical constraints of equipment and the environment constraints (delays, communication channels, bandwidth, transmission power, etc.).

8.5.3. *Complexity of computing methods*

The question of computing complexity is also an important problem. Indeed, the solutions based on deep learning developed outside the vehicular environment require significant computing and storage capacities for algorithm learning and training. This is expected to yield results at increasingly higher performances. However, the vehicles have limited computing and storage capacities and these solutions cannot be deployed in this context. This is why Zeng *et al.* (2018) propose an intrusion detection solution adapted to the capacities of vehicles. Taking into account these capacities and the development of less computing power consuming solutions are essential. Indeed, since vehicular communications require low latency, many decisions and computations will be expected to be made by the vehicles, without external support. Therefore, adaptation to these resources requires the design of low complexity computing methods, which guarantee high performance levels.

8.5.4. *Vehicle flow motion*

Although Huang *et al.* (2014), Ide *et al.* (2015) and Lv *et al.* (2015) introduce various solutions for the prediction of the vehicle flow motion, none of them has for the time being been tested in a real environment. Moreover, many improvements may be considered in order to increase performances (Huang *et al.* 2014). The prediction of vehicle flow motion is the common basis of these works, whether it is a matter of securing, controlling or managing the communications. Moreover, the development of a reliable, normalized solution, offering optimal performances, notably in terms of latency, is an essential point for the whole vehicular architecture. Moreover, the analysis of driver and vehicle user behavior may enable the development of even higher accuracy solutions. Consequently, it is an open domain for which the evaluation of proposals in a real environment is crucial.

8.6. Conclusion

The reference architecture of cooperative intelligent transportation systems has various limitations in its current state: capacity to process a significant amount of data, interoperability, guarantees of QoS, etc. This is why vehicular networks evolved toward a new paradigm: the IoV. IoV is expected to address the limitations of vehicular networks and to develop new services: driver assistance, road safety, health, etc.

The evolution of vehicular architecture and the ascent of the IoV rely on the integration of new technologies in the vehicular architecture. AI is part of these technologies. Indeed, it could offer many advantages, particularly through intelligent decision making and the automation of many services. This is why many publications have already proposed definitions of vehicular communication architectures integrating AI techniques. Nevertheless, these architectures focus essentially on an evolution of the control plane of the vehicular communication architecture and not on an evolution of other planes, namely the management and security planes, for which the integration of AI techniques could also bring many benefits.

This is why this chapter proposes the addition of a fourth plane, namely the KP, in support of all the other planes. This plane is also expected to enable multilevel decision making, a more efficient information processing and an improvement of many services. Section 8.4 of this chapter presents this KP, as well as the benefits for each plane: management (load balancing, management of distributed resources, etc.), control (motion, congestion control, dynamic routing) and security (intrusion detection, DoS, secure routing). The integration of AI in the vehicular architecture requires, nevertheless, the consideration of various points, notably the security and protection of private life, the complexity of computing methods, swarm learning and the prediction of vehicle motion.

8.7. References

Alheeti, K.M.A., Gruebler, A., and McDonald-Maier, K.D. (2015). On the detection of grey hole and rushing attacks in self-driving vehicular networks. *7th Computer Science and Electronic Engineering Conference (CEEC)*. 24–25 September 2015, Colchester, United Kingdom, 231–236.

Alheeti, K.M.A., Gruebler, A., McDonald-Maier, K.D., and Fernando, A. (2016). Prediction of DoS attacks in external communication for self-driving vehicles using a fuzzy petri net model. *IEEE International Conference on Consumer Electronics (ICCE)*. 9-11 January 2016. Las Vegas, USA, 502–503.

Aliyu, A., Abdullah, A.H., Kaiwartya, O., Cao, Y., Usman, M.J., Kumar, S., Lobiyal, D.K., and Raw, R.S. (2017). Cloud computing in VANETs: Architecture, taxonomy, and challenges. *Institution of Electronics and Telecommunication Engineers (IETE) Technical Review*, 35(5), 523–547.

Ashraf, M.I., Bennis, M., Perfecto, C., and Saad, W. (2016). Dynamic proximity-aware resource allocation in vehicle-to-vehicle (V2V) communications. *IEEE Globecom Workshops*, 1–6.

Borcoci, E., Obreja, S., and Vochin, M. (2017). Internet of vehicles functional architectures: comparative critical study. *The Ninth International Conference on Advances in Future Internet (AFIN)*. 10-14 September 2017. Rome, Italy, 10–14.

Chen, M., Tian, Y., Fortino, G., Zhang, J., and Humar, I. (2018). Cognitive Internet of vehicles. *Computer Communications*, 120, 58–70.

Clark, D.D., Partridge, C, Ramming, C., and Wroclawski, J.T. (2003). A Knowledge plane for Internet. Proceedings of the 2003 Conference on Applications, *Technologies, Architectures, and Protocols for Computer Communications*. ACM publisher. 25–29 August 2003, Karlsruhe, Germany, 3–10.

Contreras-Castillo, J., Zeadally, S., AND Guerrero Ibáñez, J.A. (2017). A seven-layered model architecture for Internet of Vehicles. *Journal of Information and Telecommunication*, 1(1), 4–22.

ETSI (2010). Intelligent Transport Systems (ITS), communications architecture. Document, European Telecommunications Standard Institute, Sophia Antipolis.

Huang, W., Song, G., Hong, H., and Xie, K. (2014). Deep architecture for traffic flow prediction: Deep belief networks with multitask learning. *IEEE Transactions on Intelligent Transportation Systems*, 15(5), 2191–2201.

Ide, C., Hadiji, F., Habel, L., Molina, A., Zaksek, T., Schreckenberg, M., Kersting, K., and Wietfeld, C. (2015). LTE connectivity and vehicular traffic prediction based on machine learning approaches. *IEEE 82nd Vehicular Technology Conference (VTC2015-Fall)*. 6–9 September 2015, Boston, USA, 1–5.

Jiacheng, C.H.E.N., Haibo, Z.H.O.U., Ning, Z., Peng, Y., Lin, G., and Xuemin, S. (2016). Software defined Internet of vehicles: Architecture, challenges and solutions. *Journal of Communications And Information Networks*, 1(1), 14–26.

Kaiwartya, O., Abdullah, A.H., Cao, Y., Altameem, A., Prasad, M., Lin, C.T., and Liu, X. (2016). Internet of vehicles: Motivation, layered architecture, network model, challenges, and future aspects. *IEEE Access*, 4, 5356–5373.

Kang, M.J. and Kang, J.W. (2016). Intrusion detection system using deep neural network for in-vehicle network security. *PLOS ONE*, 11(6), 1–17.

Krief, F. (ed.) (2010). *Communicating Embedded Networks: Network Applications*. ISTE Ltd, London, and Wiley, New York.

Krundyshev, V., Kalinin, M., and Zegzhda, P. (2018). Artificial swarm algorithm for VANET protection against routing attacks. *IEEE Industrial Cyber-Physical Systems (ICPS)*. 15 18 May 2018. St. Petersburg, Russia, 795–800.

Lai, W.K., Lin, M.T., and Yang, Y.H. (2015). A machine learning system for routing decision-making in urban vehicular ad hoc networks. *International Journal of Distributed Sensor Networks*, 11(3), 374391.

Li, J., Luo, G., Cheng, N., Yuan, Q., Wu, Z., Gao, S., and Liu, Z. (2019). An end-to-end load balancer based on deep learning for vehicular network traffic control. *IEEE Internet of Things Journal*, 6(1), 953–966.

Li, Z., Wang, C., and Jiang, C.J. (2017). User association for load balancing in vehicular networks: An online reinforcement learning approach. *IEEE Transactions on Intelligent Transportation Systems*, 18(8), 2217–2228.

Liang, L., Ye, H., and Li, G.Y. (2019). Toward intelligent vehicular networks: A machine learning framework. *IEEE Internet of Things Journal*, 6(1), 124–135.

Lv, Y., Duan, Y., Kang, W., Li, Z., and Wang, F.Y. (2015). Traffic flow prediction with big data: A deep learning approach. *IEEE Transactions on Intelligent Transportation Systems*, 16(2), 865–873.

Lyamin, N., Kleyko, D., Delooz, Q., and Vinel, A. (2018). AI-based malicious network traffic detection in VANETs. *IEEE Network*, 32(6), 15–21.

Macedo, D.F., Dos Santos, A.L., Nogueira, J.M.S., and Pujolle, G. (2008). A knowledge plane for autonomic context-aware wireless mobile ad hoc networks. *IFIP/IEEE International Conference on Management of Multimedia Networks and Services (MMNS)*. 22–26 September 2008. Samos Island, Greece, 1–13.

Mendiboure, L., Chalouf, M.A., and Krief, F. (2019a). Towards a 5G vehicular architecture. *Communication Technologies for Vehicles – Proceedings of the 14th International Workshop*. Nets4Workshops, Colmar.

Mendiboure, L., Chalouf, M.A., and Krief, F. (2019b). Edge computing based application in vehicular environments: Comparative study and main issues. *Journal of Computer Science and Technology*, 34(4), 869–886.

Movahedi, Z., Ayari, M., Langar, R., and Pujolle, G. (2012). A survey of autonomic network architectures and evaluation criteria. *IEEE Communications Surveys & Tutorials*, 14(2), 464–490.

Nyabuga, S.M., Cheruiyot, W., and Kimwele, M. (2016). Using particle swarm optimization (PSO) algorithm to protect vehicular ad hoc networks (VANETS) from denial of service (DoS) attack. *International Journal of Advanced Research in Computer Engineering & Technology (IJARCET)*, 5(3), 628–631.

Sun, Y., Song, J., Zhou, S., Guo, X., and Niu, Z. (2018). Task replication for vehicular edge computing: A combinatorial multi-armed bandit based approach. *2018 IEEE Global Communications Conference (GLOBECOM)*. 9–13 December 2018. Abu Dhabi, United Arab Emirates, 1–7.

Taherkhani, N. and Pierre, S. (2016). Centralized and localized data congestion control strategy for vehicular ad hoc networks using a machine learning clustering algorithm. *IEEE Transactions on Intelligent Transportation Systems*, 17(11), 3275–3285.

Toutouh, J. and Alba, E. (2016). Distributed fair rate congestion control for vehicular networks. *The 13th International Symposium on Distributed Computing and Artificial Intelligence (DCAI)*. 1–3 June 2016. Seville, Spain, 433–442.

Toutouh, J. and Alba, E. (2018). A swarm algorithm for collaborative traffic in vehicular networks. *Vehicular Communications*, 12, 127–137.

Xu, Y., Li, L., Soong, B.H., and Li, C. (2014). Fuzzy Q-learning based vertical handoff control for vehicular heterogeneous wireless network. *2014 IEEE International Conference on Communications (ICC)*. 10–14 June 2014, Australia.

Yao, L., Wang, J., Wang, X., Chen, A., and Wang, Y. (2018). V2X routing in a VANET based on the hidden Markov model. *IEEE Transactions on Intelligent Transportation Systems*, 19(3), 889–899.

Ye, H., Liang, L., Li, G.Y., Kim, J., Lu, L., and Wu, M. (2018). Machine learning for vehicular networks: Recent advances and application examples. *IEEE Vehicular Technology Magazine*, 13(2), 94–101.

Ye, H., Li, Y.G., and Juang, B.H.F. (2019). Deep reinforcement learning for resource allocation in V2V communications. *IEEE Transactions on Vehicular Technology*. 68(4), 3163–3173.

Zeng, Y., Qiu, M., Ming, Z., and Liu, M. (2018). Senior2Local: A machine learning based intrusion detection method for VANETs. *International Conference on Smart Computing and Communication*. 10–12 December. Tokyo, Japan, 417–426.

Zhang, G., Wu, M., Duan, W., and Huang, X. (2018a). Genetic algorithm based QoS perception routing protocol for VANETs. *Wireless Communications and Mobile Computing*. 3897857, 1–10.

Zhang, H., Bochem, A., Sun, X., and Hogrefe, D. (2018b). A security aware fuzzy enhanced reliable ant colony optimization routing in vehicular ad hoc networks. *2018 IEEE Intelligent Vehicles Symposium (IV)*. 26-30 June 2018. Changshu, China, 1071–1078.

Zheng, Q., Zheng, K., Zhang, H., and Leung, V.C. (2016). Delay-optimal virtualized radio resource scheduling in software-defined vehicular networks via stochastic learning. *IEEE Transactions on Vehicular Technology*, 65(10), 7857–7867.

PART 5

Intelligent Radio Communications

9

Artificial Intelligence Application to Cognitive Radio Networks

Badr BENMAMMAR and Asma AMRAOUI
Abou Bekr Belkaid University, Tlemcen, Algeria

9.1. Introduction

In wireless networks, field resources in terms of frequency and bandwidth availability are increasingly scarce. Consequently, new solutions are required to minimize the energy consumption and optimize the allocation of radio resources.

For flexible access to the spectrum, Mitola and Maguire (1999) introduced cognitive radio (CR) relying on software-defined radio. Software-defined radio is a radio that can realize in software form the typical functions of the radio interface generally realized in a hardware form, such as the carrier frequency, signal bandwidth and modulation. Indeed, Mitola and Maguire (1999) combined their software-defined radio experiences, as well as their passions for machine learning (ML) and artificial intelligence (AI) to set up CR technology. According to Mitola (2000), a CR is able to know, perceive and learn from its environment, and then act in order to simplify the user's life. In 2005, Haykin (2005) reviewed the concept of CR and dealt with it as brain-empowered wireless communication.

CR is a technology that detects the environment, analyzes its transmission parameters, and then makes the decisions related to resource allocation and management. However, the formulations of optimization for resource allocation offer optimal solutions sometimes at the expense of the computing time and the

Intelligent Network Management and Control,
coordinated by Badr BENMAMMAR.

processing complexity. To reduce this complexity and obtain resource allocation within reasonable, the cognitive radio network (CRN) must be equipped with learning and reasoning capacities. The cognitive engine must coordinate the CR actions using ML techniques.

Consequently, a CR must be intelligent and able to learn from its experience by interacting with its radiofrequency environment. Therefore, learning is an essential element of CR, which can be provided using AI and ML techniques. Indeed, applying AI, and in particular ML to CRN, has recently attracted considerable interest in the literature.

Learning aims to enable machines to carry out similar tasks to those of an expert. The intelligent machine senses its environment and takes measures to maximize its own utility. AI focuses on deduction, reasoning, problem solving, knowledge and learning representation (Woods 1986).

The learning process in the CRN is illustrated in Figure 9.1 and can be presented as follows (Abbas *et al.* 2015):

– detection of radiofrequency parameters such as the channel quality;

– observation of the environment and analysis of its reactions;

– learning;

– conservation of decisions and observations for model update;

– decision on resource management problems and consequent transmission error adjustment (Bkassiny *et al.* 2013; Russell 2016).

There are several works in the literature that focus on the application of AI to CRN. In Zhao and Morales-Tirado (2012), the authors introduced the use of AI and ML techniques in CR. They also presented the possible applications and the fundamental ideas of CR. In Bkassiny *et al.* (2013), the authors presented a survey on the various AI techniques, such as reinforcement learning (RL), game theory (GT), neural networks (NNs), support vector machines (SVM) and Markov models (MM). The survey discusses the strengths and weaknesses of these techniques, as well as the difficulties encountered in their applications in the CR field. In Gavrilovska *et al.* (2013), the authors studied the GT, RL and reasoning approaches such as Bayesian networks (BNs), fuzzy logic (FL) and case-based reasoning (CBR) in CRN. The survey presented by Abbas *et al.* (2015) was dedicated to FL, genetic algorithms (GA), NNs, GT, RL, SVM, CBR, decision trees (DT), BNs, MMs, multiagent systems (MASs) and artificial bee colonies algorithms.

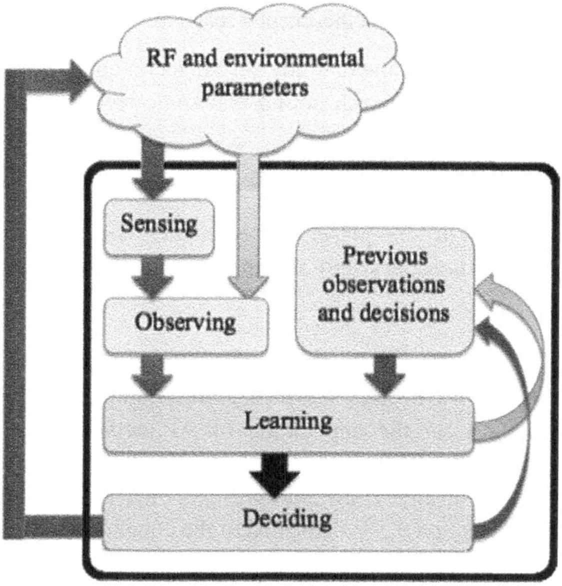

Figure 9.1. *Learning process in cognitive radio networks (Abbas* et al. *2015)*

It can however be noted that the article by Abbas *et al.* (2015) did not discuss the application of particle swarm optimization (PSO), a commonly used metaheuristic in CR. Other more recent metaheuristics that are also used in CR were ignored by Abbas *et al.* (2015), such as the firefly algorithm (FA), cuckoo search (CS) and the gravitational search algorithm (GSA).

This chapter focuses on the AI techniques that were most commonly used during the last three years in CR (the state-of-the-art of work before 2015 was already done in Abbas *et al.* (2015)).

Our focus is on work that was not discussed in previous surveys, such as the FA, CS, GSA and PSO. We also present recent work related to the application of other AI techniques in CR, namely GA, bee colony algorithms, FL, GT, NNs, MMs, SVM, CBR, DT, BNs, MASs and RL. Altogether, 17 AI techniques will be studied in this chapter.

The essential points dealt with in this chapter relate to:

– the presentation of a complete study on AI techniques, their definitions and their applications to CR. It is worth noting that some techniques have never been discussed before in this field;

– the presentation of the main tasks of CR and their corresponding challenges;

– the categorization of the presented techniques depending on the type of learning (supervised or unsupervised) and their applications depending on the CR tasks.

The rest of this chapter is organized as follows. Section 9.2 presents the cognition cycle, the main CR tasks and their corresponding challenges. Section 9.3 offers a state-of-the-art on the application of AI methods to CR. Section 9.4 proposes a categorization of the techniques presented depending on the type of learning (supervised or unsupervised) and presents their applications depending on the CR tasks. Finally, section 9.5 is dedicated to the conclusion.

9.2. Cognitive radio

9.2.1. *Cognition cycle*

A CRN follows the cognitive cycle in order to optimize its performances (see Figure 9.1). It starts with sensing the environment, then goes on with analyzing external parameters and ends by making decisions related to the allocation and dynamic management of resources, in order to improve spectral efficiency (Biglieri *et al.* 2013).

Sensing the environment: in the CR, the secondary network can use the available spectrum, but without causing interference in the primary network. Consequently, the secondary network must first detect the parameters of its environment, such as the availability of spectrum holes in the frequency.

Analysis of parameters of the environment: the detected environment parameters are used as inputs for resource management. The latter may include energy consumption minimization, interference minimization, throughput optimization, improvement of quality of service (QoS) and maximization of spectral efficiency (Wyglinski *et al.* 2009).

Decision making: in CR, decision making can rely on optimization algorithms. However, in order to reduce complexity and get a resource allocation within reasonable time, CRN use ML and AI (Qiu *et al.* 2012).

9.2.2. *CR tasks and corresponding challenges*

CR generally relies on two main tasks:

– *the "cognitive" task*, which can be obtained by spectrum detection techniques. The main challenges for these techniques are the accuracy of the decision concerning spectrum availability, the detection duration, the detection frequency and the uncertainty on the power of ambient noise, particularly at low signal-to-noise ratio (SNR) due to multipath attenuation and shadowing. To improve spectrum detection performances, cooperative detection and geolocalization technologies were proposed in the literature (Ghasemi and Sousa 2008; Wang and Liu 2011; Umar and Sheikh 2012);

– *the "reconfigurable" task*, which is used to dynamically adjust the transmission parameters in order to improve network performances. It relies on decision making, which is based on optimization algorithms. The main challenge of this task concerns the complexity and the convergence of these techniques within a limited time. This can be solved using AI and ML techniques in order to build learning models for decision making. Consequently, the choice of a learning technique for conducting a specific CR task is in itself considered a challenge.

9.3. Application of AI in CR

9.3.1. *Metaheuristics*

In the literature, metaheuristics are classified into two subcategories: those with a single solution and those that are population based. Single solution metaheuristics are iterative approaches that start with a single initial solution and improve it with each iteration taking advantage of its neighborhood. Population-based metaheuristics explore the research space using a set of solutions known as "population". This latter category is also divided into two subcategories: evolutionary algorithms and algorithms based on swarm intelligence.

Evolutionary algorithms draw their inspiration from the natural evolution of species, and more precisely from the natural selection principle stated in the theory of evolution, developed by Darwin (2009) in his work entitled *On the Origin of Species: By Means of Natural Selection or the Preservation of Favored Races in the Struggle for Life*. The algorithms based on swarm intelligence were introduced for the first time by Gerardo and Wang (1993) in their work entitled "Swarm intelligence in cellular robotics systems", which describes the behavior of a group of robots cooperating to accomplish a task or solve a problem.

The following section presents the application of six metaheuristics most commonly used in CRN, namely FA, CS, bee colony algorithms, GA, GSA and PSO.

9.3.1.1. *Firefly algorithm*

The FA is an optimization approach based on swarm intelligence that was proposed by Xin-She Yang in 2008 (Surafel and Ngnotchouye 2017). Its principle is inspired by the lighting behavior of fireflies. Indeed, each potential solution is assimilated with a firefly, whose luminosity is proportionally related to its quality (quality of the solution). The following works dealt with the application of FA to Cognitive Radio Network (CRN).

The FA was adapted by the authors in a CRN based on orthogonal frequency-division multiplexing (OFDM) (Saoucha and Benmammar 2017). Multiobjective optimization was also used to optimize the quality of communication between secondary users (SU). The performances of the FA were validated through a comparison with the PSO algorithm and with the cross entropy in terms of convergence speed, quality of the solution and stability.

The FA was used in Tounsi and Babes (2017) to solve the problem of power control and channel allocation in CRN. A modified version of the FA, using the new attractiveness factor, was proposed to solve this problem. A theoretical analysis was presented in this article to prove the efficiency and the existence of Nash equilibrium concerning the proposed strategy. The results presented in this article show that the proposed method outperforms the approaches in the literature in terms of convergence speed.

To estimate the amplitudes of cancellation subcarriers, the authors in Elahi *et al.* (2017) proposed two search algorithms: the GA and the FA. According to the simulation results, the proposed algorithms enable a better reduction of the secondary lobes compared to the techniques currently mentioned by the literature.

In Ghanem *et al.* (2016), the problem of attack based on primary user emulation (PUE) is solved by means of a localization defense model based on the use of the FA. The CR users cooperate to detect and localize the attacker comparing its location with the position of the primary users (PU). The simulation results are compared with existing methods and prove that the FA reduces the localization error and requires fewer SUs to cooperate.

9.3.1.2. *Cuckoo search*

CS is an optimization approach based on swarm intelligence, which was proposed by Yang and Deb (2017). Its development draws inspiration from the parasitic behavior of certain bird species. In the real world, cuckoos lay their eggs in the nests of other bird species. In most cases, the host bird believes that the laid eggs are hers and, consequently, takes care of them. However, in certain cases the eggs laid are discovered and discarded by the host bird or the host bird leaves the nest.

The following works focus on the application of CS in the CRN.

A new method for the OFDM-based estimation of the channel state in a CRN was proposed in Manjith (2016). The method is hybridization between *bacterial foraging optimization* (BFO) and a modified CS algorithm.

In Kaur *et al.* (2018a), the authors proposed a new multiuser CR system, as well as its optimization by means of CS algorithm. The transmission parameters of several SUs are considered according to the IEEE 802.22 WRAN standard. The optimization results were compared to another efficient optimization technique based on biogeography and simulated annealing.

9.3.1.3. *Bee colony algorithm*

Various groups of researchers independently participated in the development of bee algorithms during the last 10 years. Tovey, at Georgia Tech, in collaboration with Nakrani, at the University of Oxford, proposed the honey bee algorithm for the first time in 2004 (Nakrani and Tovey 2004). The virtual bee algorithm was created in 2005 by Xin-She Yang at the University of Cambridge to solve numerical optimization problems (Yang 2005). Haddad *et al.* (2006) presented the honey bees mating optimization algorithm.

The artificial bee colony (ABC) algorithm was developed in 2005 by Karaboga (2005) for the optimization of numerical functions.

The following works focused on the application of bee colony algorithms in CRN.

A new spectral handover algorithm based on the ABC algorithm in a CRN was proposed in Bayrakdar and Calhan (2018). In the algorithm proposed by the authors, the spectrum availability characteristic is observed on the basis of bee missions in order to minimize the spectral handover delay and maximize the probability of finding an inactive channel. The main advantage of this algorithm is that the delay of the spectral handover of SUs is considerably reduced for a different number of users, without reducing the probability of finding an available channel.

A hybrid algorithm between artificial bee colonies and GA was proposed in Elghamrawy (2018) to optimize the use of spectrum by detecting the PUE attacks and increasing the detection probability. The proposed algorithm integrates the genetic operators with the ABC algorithm in order to reach equilibrium between use and exploration to find the optimal solution. The simulation results indicate promising performances of the proposed algorithm for spectrum detection optimization, compared to recent detection algorithms.

The main objective of Zaheer *et al.* (2016) is to minimize the transmission powers and thus reduce the interferences in a CRN by using the ABC algorithm.

9.3.1.4. *Genetic algorithms*

The principle of GA was introduced by John Holland at the University of Michigan in the United States, in the 1960s (Holland 1992), and highlighted by the reference work by Goldberg (1989). In a genetic algorithm, a population is constituted by a set of individuals, each of which is identified by a set of genes known as "chromosomes". Reproduction involves the recombination of chromosomes of two primary individuals, thus giving birth to child individuals whose genetic fingerprint is inherited from parents. However, the genetic code of the children may contain genes that their parents do not have, thus modeling the mutation genetic phenomenon. The latter enables changes in the morphology of species, always leading to a better adaptation to the natural environment.

The following works focus on the application of GA to CRN.

The problem formulation, the development and use of a genetic algorithm for channel assignment in a CRN was presented by Elhachmi and Guennoun (2016). This approach offers the PU and SU an efficient means to access the available spectrum. Compared to existing methods, simulation results prove that the proposed algorithm yields satisfactory results in terms of interferences and throughput.

In Jiao and Joe (2016), the authors considered a new model of CRN in which the networks of PU are constituted of heterogeneous PU. The authors consider the problem of energy-efficient resource allocation for the CR user having a coverage area in which heterogeneous PU operate simultaneously via multiradio access technology. The authors proposed a research diagram based on two level crossover GA to obtain an optimal solution in terms of power and throughput. Simulation results show that the algorithm proposed by the authors is stable and its convergence is more rapid.

9.3.1.5. *Gravitational search algorithm*

In 2009, Rashedi *et al.* (2009) developed a GSA, which is an optimization metaheuristics inspired by nature. GSA relies on Newton's law of gravitation, according to which gravitation is an attraction between massive bodies. The masses of the bodies (solutions) are proportional to their values of objective functions (costs). With each iteration, masses are mutually attracted by gravitational forces. The heaviest mass exerts the strongest force of attraction. Consequently, the heaviest masses, which are probably close to the global optimum, attract other masses depending on their distances. Each object is determined by four specifications: position, inertia, active gravitational mass and passive gravitational mass. The position corresponds to a solution to the problem; inertia and gravitational masses are determined using the objective function.

The following works focused on the application of the GSA in CRN.

Guo *et al.* (2018) present a new method for solving the problem of spectrum waste in CRN. This method relies on graph coloring and GSA. The authors compared the performances of their algorithm with the PSO and GA.

In Kaur *et al.* (2018b), a hybrid algorithm of PSO and gravitational research is presented for CRN optimization. A new CR environment is proposed, enabling several SUs to access the spectrum while their channels undergo Nakagami-m fading. The transmission factors belonging to several SUs and relying on IEEE 802.22 WRAN standard are optimized to reach several objectives related to the expected QoS using PSO, GSA and the hybrid PSOGSA algorithm. Objective functions that are modified and influenced by fading are established for the optimization task. The optimization results indicate an improved performance of the hybrid algorithm compared to other basic techniques.

9.3.1.6. *Particle swarm optimization*

PSO is an optimization metaheuristics in the family of evolutionary algorithms and proposed by Russel Eberhart (electrical engineer) and James Kennedy (sociopsychologist) (Kennedy and Eberhart 1995). PSO has its source in the observations made during computer simulations of flocks of birds and fish schools (Craig 1987; Heppner and Grenander 1990). Indeed, the PSO draws heavily on the observation of gregarious relations of migratory birds, which, in order to travel "long distances" (migration, search for food, aerial displays, etc.), must optimize their motions in terms of consumed energy and time (etc.), as for example in the V-shaped flight formation presented in Figure 9.2.

Figure 9.2. *V-shaped formation of Anser flight (Bestaoui 2015)*

The following work focuses on the application of PSO in CRN.

A technique based on PSO and on the received signal strength indicator for the detection of the position of PU and the PUE–based attacker was proposed in Fihri *et al.* (2018). The authors aim to improve the detection accuracy and reduce the risk of false alarms.

The effect of eigenvalues of the covariance matrix on samples received through the SNR estimation method was analyzed in Manesh *et al.* (2017). The authors proposed the use of PSO in the SNR estimation technique based on eigenvalues in order to optimize these parameters. The results of the proposed method are compared to those of the original SNR estimation method and the results validate the improvement obtained by the proposed technique compared to the original technique.

The simultaneous wireless information and the multiple user power transfer for the CRN relying on PSO and on semidefinite relaxation were studied in Tuan and Koo (2017). A secondary emitter with an antenna array provides information and energy to several secondary receivers with a single antenna. The authors proved through simulations that their algorithm features rapid convergence and better performances compared to other existing systems.

In Zhai and Wang (2017), the authors used PSO to solve the crowdsourcing paradigm, according to which mobile users are assigned the task of spectrum detection. Simulation results show that the proposed algorithm reaches higher performances compared to those of other algorithms.

In Tang and Xin (2016), the authors used PSO to study the compromise between utility and energy consumption in a CRN based on OFDM. Considering the low convergence of the original PSO around local optima, an enhanced version combining chaos theory is proposed in this study in order to help PSO identify solutions around the best global results. Using simulations, the authors proved that their algorithm requires a smaller number of iterations and can reach a higher energy efficiency than the other algorithms.

Tuan and Koo (2016) proposed a hybrid method based on PSO and Brute-Force Search (BFS). This method is used for the maximization of the SU throughput in a full-duplex CRN, when it has two distinct antennas and a capacity to autoeliminate its interferences. The simulations show that for certain values of the parameters, the considered system provides a much higher throughput than the previously proposed systems.

Alhammadi *et al.* (2016) discussed the three spectral handover mechanisms (proactive, reactive and hybrid) used to reduce the handover delay. The article contains an implementation of the PSO algorithm to minimize the total service time of the spectral handover to optimal value. The numerical results show that PSO significantly reduces the total service time compared to otter spectral handover systems.

9.3.2. *Fuzzy logic*

In 1965, relying on mathematical theory on fuzzy sets, Lotfi Zadeh developed FL, which is an extension of Boolean logic (Zadeh 1965). FL makes it possible to consider imprecisions and uncertainties, due to which the reasoning involving it is rendered significantly more flexible. The following work focuses on the application of FL in CRN.

Banerjee *et al.* (2017) proposed a new decision making method based on FL for the relay selection, unlike many existing works in which the signal to interference plus noise ratio (SINR) is considered the only parameter for relay selection. To find the best relays, the authors conducted a broad simulation study. The simulation results reveal the influence of various parameters on the selection of the best relay.

9.3.3. *Game theory*

The first known discussion on GT was mentioned in a letter written by James Waldegrave in 1713. The GT is used as a decision making technique, where several players must make choices and consequently influence the interests of other players. Each player decides its actions depending on the history of the actions selected by other players during the previous rounds of the game.

In CRN, the nodes are the game players and the actions are the parameters of the radio environment, such as the emission power and the channel selection. These actions are made by the nodes on the basis of observations represented by the parameters of the environment, such as channel availability, channel quality and interferences. Consequently, each node draws lessons from its past actions, observes the actions of other nodes and consequently modifies its actions (Bellhouse 2017). The following works focus on the application of GT in CRN.

An approach based on the GT using the Stackelberg game to secure a network of CR sensors against the attack involving spectrum detection data falsification was proposed in Abdalzaher *et al.* (2017); the purpose of this attack was to corrupt the spectrum decisions communicated by the sensor nodes to the merging center by imposing an interference power. Simulations indicate the improvement of the performances of the proposed protection model compared to the two basic defense mechanisms, namely random defense mechanisms and those whose protection is equal to static signal to noise ratio.

The problems related to the security of the physical layer and to energy efficiency because of power control and relay cooperation, where decode-and-forward and amplify-and-forward protocols are considered, were studied in Fang *et al.* (2017). The authors proposed a Stackelberg game model with one leader and one follower in the presence of multiple listeners, in which an optimal strategy of power allocation and pricing can be determined in order to maximize player utility. The simulations conducted by the authors prove that the proposed game model improves the network energy efficiency and offers better performances against eavesdropping attacks, compared to Nash equilibrium systems, rand and direct transmission.

In Roy *et al.* (2017), the authors used the GT to study the conflict and the cooperation between two levels of SUs (real time and not real time). A model of the auction game is proposed in order to analyzer the decision-making process and efficiently allocate an inactive channel to a pair of SUs (real time and not real time) belonging to a group of users.

A two-level Stackelberg game model, in which the PU and SU act, respectively, as leaders and followers in order to improve the energy efficiency of nodes in a multiple jumps CRN was proposed in Shu *et al.* (2016). The simulations conducted proved the relevance of the authors' proposals.

9.3.4. *Neural networks*

NNs were introduced by Warren McCulloch and Walter Pitts in 1943 and draw their inspiration from the central nervous system. Similarly to biological NNs, an artificial NN is formed of nodes, equally known as "neurons" or "processing elements", which are connected together to form a network.

The artificial NN receives information from all neighboring neurons and provides an output depending on its weight and on the activation functions. The adaptive weights can represent connection forces between neurons. To accomplish the learning process, the weights must be adjusted until the network output is approximately equal to the desired output.

Artificial NNs were used to enable CR to learn from the environment and make decisions in order to improve the QoS of the communication system (Haykin 2008; Rojas 2013). The following works focus on the application of NNs to CRN.

Supraja and Pitchai (2019) presented a hybrid system constituted of a genetic algorithm, PSO and a back-propagation NN as supervised learning algorithm enabling the prediction of spectrum profiles in CRN.

Zhang *et al.* (2017) used the convolutional NN in an automatic system for the recognition of CR wave shapes. The proposed system can identify up to eight types of signals. The classification results were proven by the authors through simulations.

Liu *et al.* (2019) studied the compromise between energy efficiency and spectral efficiency for the PU and SU in a CRN. A feed-forward NN is designed and a back-propagation analog algorithm is developed to learn the optimal parameters of the algorithm proposed by the authors. Simulations are provided to confirm the efficiency of the proposed algorithm.

9.3.5. *Markov models*

MM is used for modeling random processes passing from one state to another in time. The random process is without memory where the future states depend only on the present state (Norris 1998; Ching and Ng 2006). In MMs, the states are visible to the observer; nevertheless, in the hidden Markov model (HMM), certain states are masked or are not explicitly visible (Fraser 2008). The following works focus on the application of MMs to CRN.

To solve the problems of power distributed control in a network of wireless cognitive sensors, a power control mechanism based on the HMM is proposed in Zhu *et al.* (2017) depending on the difference and on the independence of channel sensing results among the network users. The simulations indicate that, besides improving the energy efficiency, the power control mechanism based on the HMM model better reaches the SINR target compared to other methods.

An adaptive method for double threshold energy detection based on the MM was proposed in Liu *et al.* (2017). With the use of this method, the modified Markov method takes into account the time-variable characteristic of the channel occupation to solve the state of the channel "under confusion". The simulations show that the proposed method yields better performances in terms of detection compared to other existing methods.

A method to build a radio environment map (REM) in an environment with several PU was proposed in Ichikawa and Fujii (2017). REM provides statistical information on the activity of the PU at each location. It also enables the SU to dynamically access the licensed band. Simulations show that the proposed method has better performances than the existing unsupervised classification method.

9.3.6. *Support vector machines*

SVMs are an ML approach that use a non-probabilistic linear classifier for data classification into two categories. The following works focus on the application of SVMs in CRN.

Four supervised ML techniques were used to study the prediction of PU activity in Agarwal *et al.* (2016). Two of these originate in NNs and the other two in SVMs. The results highlight the analysis of learning techniques depending on various traffic statistics and suggest the best learning model enabling the accurate prediction of the PU.

A small dimension probability vector for the cooperative spectrum detection based on ML techniques in a CRN was proposed in Lu *et al.* (2016). The K-means classification algorithm, the SVM and the probability vector were studied by the authors. Considering a CRN with 1 PU and N SUs, the proposed probability may reduce the dimension of the existing energy vector from N dimensions to two dimensions, driving similar or even better detection accuracy, a shorter learning duration and a shorter classification time.

9.3.7. *Case-based reasoning*

CBR is a problem solving paradigm based on the reuse of past experiences to solve new problems. CBR builds a database with information on past situations, problems, solutions and their advantages. The new problems are then solved by finding the most similar case in the memory and deducing the solution to the current situation (Kolodner 2014). The following work focuses on the application of CBR to CRN.

A Q-learning method, based on the case of dynamic access to the spectrum improving and stabilizing the performances of cognitive cellular systems with dynamic topologies, was proposed in Morozs *et al.* (2016). The proposed approach is a combination of classical distributed Q-learning and a new implementation of CBR algorithm seeking to facilitate a certain number of learning processes executed in parallel. Simulations show that the proposed case-based Q-learning approach enables constant improvement of the QoS under conditions of dynamic and asymmetric network topology and with traffic load.

9.3.8. *Decision trees*

DTs are one of the major data structures of ML. Their operation relies on heuristics that, while satisfying intuition, yield outstanding results in practice. Due to their tree structure, DTs are readable by the human being, unlike other approaches in which the classifier builds a "black box", as in the case of NNs. The following work focuses on the application of DTs to CRN.

A new algorithm combining the random forest (several DTs) to reduce interferences was proposed in Wang and Yang (2016a). This advancement also enables the significant improvement in network throughput.

A new spectrum detection diagram relying on DTs for the classification of the MAC layer protocol was proposed in Wang and Yang (2016b). The simulations confirm that the new proposed method could significantly improve the network throughput.

9.3.9. *Bayesian networks*

BNs are graphical probabilistic models relying on the interaction between various nodes to learn and starting from each node involved in the process. BNs play a role in the decision making process if they are associated with other tools in order to form influence diagrams (Bolstad 2004).

The following work focuses on the application of BNs in CRN.

A BN-based model for dealing with modularity and uncertainty was proposed in Elderini *et al.* (2017). The Bayesian model enables the qualitative and quantitative addition of parameters that influence the fault probability and the SINR in a CRN.

In Salahdine *et al.* (2017a), BNs are considered among the uncertainty management techniques in CRN. A use case involves CRN modeling by a graph in which each node represents a SU and each edge represents the communication link between corresponding nodes.

A hybrid method between a Bayesian model and a trilateration technique, which is used to obtain a good approximation of the position of the PUE attacker, was proposed in Fihri *et al.* (2017). The Bayesian decision theory based on the loss function and the conditional probability makes it possible to determine the existence of the PUE attacker in the uncertainty area.

An efficient and rapid sensing method, in which the Toeplitz sensing matrix and the Bayesian model are combined to deal with uncertainties and reduce the random character of measures, was proposed in Salahdine *et al.* (2017b). The proposed method was thoroughly implemented and tested, yielding satisfactory results.

A nonparametric Bayesian approach to the clustering of subchannels in CRN based on OFDMA was proposed in Ahmed *et al.* (2017). The approach uses the traffic functionalities of each subchannel to obtain statistics on its idle/busy period. Based on the harvested energy, the SU determines the energy detection threshold so that it can maximize its rate of spectrum use while minimizing the interferences with the PU.

9.3.10. *MASs and RL*

Jacques Ferber presented the MAS as an intelligent entity conscious of its environment, capable of skillfully acting and autonomously communicating. They contain the environment, objects, agents and various relations between these entities

(Ferber and Weiss 1999). With MAS, users can interact, negotiate and cooperate to enable more efficient communication between the entities in the network.

Their use in CRN enables users to manage their own spectrum in a dynamic and decentralized manner. The agents sense their environment and consequently react. The association of MAS and CR enables a better use of the unused spectrum and a optimal management of radio resources, while reducing the risk of interferences.

On the other hand, RL is a domain of ML that makes it possible to solve sequential decision problems under uncertainty. It plays an essential role with SM, as they enable agents to discover the situation and take measures by trial and error, in order to maximize the cumulated reward. In RL, an agent must take into account the immediate advantages and consequences of its actions to optimize the long-term system performances (Wiering and Van Otterlo 2012).

Deep reinforcement learning (DRL) uses deep learning and RL principles to create efficient algorithms that can be applied to fields such as robotics, video games, finance and health care (François-Lavet *et al.* 2018). By implementing a deep learning architecture (deep NNs, etc.) with a RL algorithm (Q-learning, etc.), it is possible to create a powerful DRL model that can adapt previously insoluble problems. The following work focuses on the application of MASs, RL and DRL in CRN.

The optimization of opportunistic accessibility of channels in railway CR environments was studied in Yin *et al.* (2017). The model proposed by the authors involves a Bayesian inference enabling the calculation of the probability of successful transmission on a single station, as well as a team cooperation aiming to optimize network performances in a group of base stations.

An efficient transmission mode based on the Q-learning algorithm was proposed in cooperative CRN in Rahman *et al.* (2016). The state, the action and the reward are defined to obtain good performances in terms of delay and energy efficiency during data transmission, as well as for the interference caused to PU during transmissions achieved by the SU. The simulations show that the diagram proposed by the authors can efficiently assume the determination of the transmission mode and surpasses the classical diagrams mentioned in the literature.

The efficient allocation of transmission power between the SU, without generating interference for the PU, is the objective of the authors in Lall *et al.* (2016). Three mixed strategies (correlated equilibrium) were used to control the transmission power during learning. The experimental results indicated that the proposed algorithm by far surpasses its classical counterparts.

The routing of several flows generated by the SU toward a given destination, considering the PU presence, was approached by Pourpeighambar *et al.* (2017). Each SU is expected to egoistically minimize the end-to-end delay of its flow, while meeting the QoS demands of the PU. For a rapid adaptation of the SU routing decision to environment changes and to their non-cooperative interaction, the authors formulated the routing problem as a stochastic learning process represented by non-cooperative games. Then they proposed a system based on swarm RL to solve the routing problem and thus avoid the information exchanges between competing SU. The simulations show that the proposed diagram converges in a demonstrable manner and prove its efficiency for reducing the delay while meeting the QoS demands of the PU.

An algorithm based on RL for the management of power assignment for the transmission channel and the control channel in CRN was proposed in Lin *et al.* (2016). The simulation results show that this new algorithm brings a significant improvement in terms of compromise between the control channel reliability and the transmission channel efficiency.

An algorithm for selecting a channel for data transmission and predicting the duration it will remain idle for in order to minimize the time dedicated to its detection was proposed in Raj *et al.* (2018). This algorithm involves two stages: an RL approach for the selection of channels and a Bayesian approach for determining the duration for which detection can be ignored.

The spectrum sharing in a CRN composed of a PU and SU was studied in (Xingjian *et al.* 2018). The PU and the SU work in a non-cooperative manner. The authors aim to develop a power control method based on learning for the SU in order to share the common spectrum with the PU. The authors developed a DRL-based method, which the SU can use to intelligently adjust its emission power, so that after two cycles of interaction with the PU, the two users can send their own data with the required QoS.

9.4. Categorization and use of techniques in CR

Data availability influences the choice of the learning technique to be used. Supervised learning is used when learning data are labeled and the CR has previous information on the environment. Unsupervised algorithms do not require labeled learning data. Unsupervised learning is used when certain RF components are not known by the CRN, which enables it to operate autonomously with no previous knowledge.

There are differences between learning techniques in terms of strengths, limitations, challenges and applications in CRN.

DTs, NNs, SVMs and CBR are considered supervised learning techniques. The GT, RL, MMs, metaheuristics, FL and BNs are considered unsupervised learning techniques.

It is worth noting that all the techniques studied in this chapter are used both for sensing and for decision in CR, except for the metaheuristics and for the CBR, which are used exclusively for decision. DTs, GT and FL should be used for sensing taking into account the capacity of the spectrum detection technique employed.

MAS can be used in combination with all the techniques presented in this chapter.

9.5. Conclusion

This chapter presents a complete study of the AI techniques used in CRN. The definitions of various techniques and their applications in CR are also discussed in this work. Indeed, a complete state-of-the-art on the application of the FA, CS, GSA, PSO, GA, bee colony algorithms, FL, GT, NNs, MMs, SVMs, CBR, DTs, BNs, MASs and RL in CRNs was presented in this chapter. It also presented the main CR tasks and their corresponding challenges, classified the presented techniques depending on the type of learning (supervised or unsupervised) and presented their applications depending on two CR tasks (sensing and decision).

9.6. References

Abbas, N., Nasser, Y., and El Ahmad, K. (2015). Recent advances on artificial intelligence and learning techniques in cognitive radio networks. *EURASIP Journal on Wireless Communications and Networking (JWCN)*, 174 (2015), 1–20.

Abdalzaher, M.S., Seddik, K., and Muta, O. (2017). Using Stackelberg game to enhance cognitive radio sensor networks security. *IET Communications*, 11(9), 1503–1511.

Agarwal, A., Dubey, S., Khan, M. A., Gangopadhyay, R., and Debnath, S. (2016). Learning based primary user activity prediction in cognitive radio networks for efficient dynamic spectrum access. *International Conference on Signal Processing and Communications*. 12–15 June 2016, Bangalore, India, 1–5.

Ahmed, M.E., Kim, D.I., Kim, J.Y., and Shin, Y. (2017). Energy-arrival-aware detection threshold in wireless-powered cognitive radio networks. *IEEE Transactions on Vehicular Technology*, 66(10), 9201– 9213.

Alhammadi, A., Roslee, M., and Yusoff Alias, M. (2016). Analysis of spectrum handoff schemes in cognitive radio network using particle swarm optimization. *3rd International Symposium on Telecommunication Technologies*. 28–30 November 2016, Kuala Lumpur, Malaysia, 103–107.

Banerjee, J.S., Chakraborty, A., and Chattopadhyay, A. (2017). Fuzzy based relay selection for secondary transmission in cooperative cognitive radio networks. *Advances in Optical Science and Engineering*. 279–287.

Bayrakdar, M.E. and Çalhan, A. (2018). Artificial bee colony–based spectrum handoff algorithm in wireless cognitive radio networks. *International Journal of Communication Systems (IJCS)*, 31(5), 1–16.

Bellhouse, D. (2007). The problem of Waldegrave. *Electronic Journal for the History of Probability and Statistics*, 3(2), 1–12.

Beni, G., Wang, J. (1993). Swarm Intelligence in Cellular Robotic Systems. In *Robots and Biological Systems: Towards a New Bionics?*, Dario P., Sandini G., Aebischer P. (eds). Springer, Berlin, Heidelberg.

Bestaoui, A.A. (2015). Gestion de spectre dans un réseau de radio cognitive en utilisant l'algorithme d'optimisation par essaim de particules. Master's thesis, University of Tlemcen, Algeria.

Biglieri, E., Goldsmith, A.J., Greenstein, L.J., Mandayam, N.B., and Poor, H.V. (2013). *Principles of Cognitive Radio*. Cambridge University Press, Cambridge.

Bkassiny, M., Li, Y., and Jayaweera, S.K. (2013). A survey on machine-learning techniques in cognitive radios. *IEEE Communications Surveys & Tutorials*, 15(3), 1136–1159.

Bolstad, W.M. (2004). *Introduction to Bayesian Statistics*. John Wiley & Sons, Hoboken.

Ching, W.-K. and Ng, M.K. (2006). *Markov Chains. Models, Algorithms and Applications*. Springer, Berlin.

Craig, W.R. (1987). Flocks, herds, and schools: A distributed behavioral model. *Computer Graphics*, 21(4), 25–34.

Darwin, C. (2009). *The Origin of Species: By Means of Natural Selection, or the Preservation of Favoured Races in the Struggle for Life*. Cambridge University Press, Cambridge.

Elahi, A., Qureshi, I.M., Atif, M., and Gul, N. (2017). Interference reduction in Cognitive radio networks using genetic and firefly algorithms. *International Conference on Communication, Computing and Digital Systems*.

Elderini, T., Kaabouch, N., and Reyes, H. (2017). Outage probability estimation technique based on a Bayesian model for cognitive radio networks. *IEEE 7th Annual Computing and Communication Workshop and Conference*.

Elghamrawy, S.M. (2018). Security in cognitive radio network: Defense against primary user emulation attacks using genetic artificial bee colony (GABC) algorithm. *Future Generation Computer Systems (FGCS)*. 85(8), 1–19.

Elhachmi, J. and Guennoun, Z. (2016). Cognitive radio spectrum allocation using genetic algorithm. *EURASIP Journal on Wireless Communications and Networking*, 133 (2016), 1–11.

Fang, H., Xu, L., and Raymond Choo, K.-K. (2017). Stackelberg game based relay selection for physical layer security and energy efficiency enhancement in cognitive radio networks. *Applied Mathematics and Computation*, 296, 153–167.

Ferber, J. and Weiss, G. (1999). *Multiagent Systems: An Introduction to Distributed Artificial Intelligence*. Addison-Wesley, Boston.

Fihri, W.F., Arjoune, Y., El Ghazi, H., Kaabouch, N., and El Majd, B.A. (2017). Bayesian decision model with trilateration for primary user emulation attack localization in cognitive radio networks. *International Symposium on Networks, Computers and Communications*. 16–17 May 2017. Marrakech, Morocco, 1–6.

Fihri, W.F., Arjoune, Y., El Ghazi, H., Kaabouch, N., and El Majd, B.A. (2018). A particle swarm optimization based algorithm for primary user emulation attack detection. *IEEE 8th Annual Computing and Communication Workshop and Conference*. 8–10 January 2018, Las Vegas, USA, 823–827.

François-Lavet, V., Henderson, P., Islam, R., Bellemare, M.G., and Pineau, J. An introduction to deep reinforcement learning. *Foundations and Trends® in Machine Learning*, 11(3/4), 219–354.

Fraser, A.M. (2008). *Hidden Markov Models and Dynamical Systems*. Siam, Philadelphia.

Gavrilovska, L., Atanasovski, V., Macaluso, I., and Da Silva, L. (2013). Learning and reasoning in cognitive radio networks. *IEEE Communications Surveys Tutor*, 15(4), 1761–1777.

Ghanem, W.R., Shokair, M., and Desouky, M.I. (2016). An improved primary user emulation attack detection in cognitive radio networks based on firefly optimization algorithm. *33rd National Radio Science Conference*. 22–25 February 2016, Aswan, Egypt, 178–187.

Ghasemi, A. and Sousa, E.S. (2008). Spectrum sensing in cognitive radio networks: Requirements, challenges and design trade-offs. *IEEE Communications Magazine*, 46(4), 32–39.

Goldberg, D.E. (1989). *Genetic Algorithms in Search, Optimization and Machine Learning*. Addison-Wesley Longman Publishing, Boston.

Guo, L., Chen, Z., and Huang, L. (2018). A novel cognitive radio spectrum allocation scheme with chaotic gravitational search algorithm. *International Journal of Embedded Systems*, 10(2), 161–167.

Haddad, O.B., Afshar, A., and Mariño, M.A. (2006). Honey-bees mating optimization (HBMO) algorithm: A new heuristic approach for water resources optimization. *Water Resources Management*, 20(5), 661–680.

Haykin, S. (2005). Cognitive radio: Brain-empowered wireless communications. *IEEE Journal on Selected Areas in Communications*, 23(2), 201–220.

Haykin, S. (2008). *Neural Networks and Learning Machines*. Pearson, London.

Heppner, F. and Grenander, U. (1990). *A Stochastic Nonlinear Model for Coordinated Bird Flocks*. AAAS Publication, Washington.

Holland, J.H. (1992). *Adaptation in Natural and Artificial Systems: An Introductory Analysis with Applications to Biology, Control, and Artificial Intelligence*. MIT Press, Cambridge.

Ichikawa, K. and Fujii, T. (2017). Radio environment map construction using hidden Markov model in multiple primary user environment. *International Conference on Computing, Networking and Communications*. 26–29 January 2017. Santa Clara, CA, USA, 272–276.

Jiao, Y. and Joe, I. (2016). Energy-efficient resource allocation for heterogeneous cognitive radio network based on two-tier crossover genetic algorithm. *Journal of Communications and Networks*, 18(1), 112–122.

Karaboga, D. (2005). An idea based on honey bee swarm for numerical optimization, volume 200. Technical report, Erciyes University.

Kaur, K., Rattan, M., and Singh Patterh, M. (2018a). Cuckoo search based optimization of multiuser cognitive radio system under the effect of shadowing. *Wireless Personal Communications*, 99(3), 1217–1230.

Kaur, K., Rattan, M., and Singh Patterh, M. (2018b). Cognitive radio design optimization over fading channels using PSO, GSA and hybrid PSOGSA. *Second International Conference on Intelligent Computing and Control Systems*. 14–15 June 2018, Madurai, India, 1700–1706.

Kennedy, J. and Eberhart, R.C. (1995). Particle swarm optimization. *Proceedings of the IEEE International Conference on Neural Networks IV*. 27 November–1 December 1995, University of Western Australia, Perth, 1942–1948.

Kolodner, J. (2014). *Case-Based Reasoning*. Morgan Kaufmann, Burlington.

Lall, S., Sadhu, A.K., Konar, A., Mallik, K.K., and Ghosh, S. (2016). Multiagent reinfocement learning for stochastic power management in cognitive radio network. *International Conference on Microelectronics, Computing and Communications*. 23 January 2016, Durgapur, India, 1–6.

Lin, Y., Wang, C., Wang, J., and Dou, Z. (2016). A novel dynamic spectrum access framework based on reinforcement learning for cognitive radio sensor networks. *Sensors*, 16(10), 1675.

Liu, Y., Liang, J., Xiao, N., Yuan, X., Zhang, Z., Hu, M., and Hu, Y. (2017). Adaptive double threshold energy detection based on Markov model for cognitive radio. *PLOS ONE*, 12(5), 1–18.

Liu, M., Song, T., Hu, J., Yang, J., and Gui, G. (2019). Deep learning-inspired message passing algorithm for efficient resource allocation in cognitive radio networks. *IEEE Transactions on Vehicular Technology*, 68(1), 641–665.

Lu, Y., Zhu, P., Wang, D., and Fattouche, M. (2016). Machine learning techniques with probability vector for cooperative spectrum sensing in cognitive radio networks. *IEEE Wireless Communications and Networking Conference.* April 3-6 2016, Doha, Qatar.

Manesh, M.R., Quadri, A., Subramaniam, S., and Kaabouch, N. (2017). An optimized SNR estimation technique using particle swarm optimization algorithm. *IEEE 7th Annual Computing and Communication Workshop and Conference.* 9–11 January 2017, Las Vegas, USA, 1–6.

Manjith, R. (2016). A hybrid of BFO and MCS algorithms for channel estimation of cognitive radio system. *Arabian Journal for Science and Engineering*, 41(3), 841–852.

Mitola, J. (2000). Cognitive radio: an integrated agent architecture for software defined radio. PhD Thesis, Royal Institute of Technology, Stockholm.

Mitola, J. and Maguire, G.Q. (1999). Cognitive radio: Making software radios more personal. *IEEE Personal Communications*, 6(4), 13–18.

Morozs, N., Clarke, T., and Grace, D. (2016). Cognitive spectrum management in dynamic cellular environments: A case-based Q-learning approach. *Engineering Applications of Artificial Intelligence*, 55, 239–249.

Nakrani, S. and Tovey, C. (2004). On honey bees and dynamic server allocation in internet hosting centers. *Adaptive Behavior*, 12(3–4), 223–240.

Norris, J.R. (1998). *Markov Chains.* Cambridge University Press, Cambridge.

Pourpeighambar, B., Dehghan, M., and Sabaei, M. (2017). Non-cooperative reinforcement learning based routing in cognitive radio networks. *Computer Communications*, 106, 11–23.

Qiu, R. C., Hu, Z., Li, H., and Wicks, M. C. (2012) *Cognitive Radio Communication and Networking: Principles and Practice.* John Wiley & Sons, Hoboken.

Rahman, M.A., Lee, Y.-D., and Koo, I. (2016). An efficient transmission mode selection based on reinforcement learning for cooperative cognitive radio networks. *Human-centric Computing and Information Sciences*, 6(1), 2.

Raj, V., Dias, I., Tholeti, T., and Kalyani, S. (2018). Spectrum access in cognitive radio using a two-stage reinforcement learning approach. *IEEE Journal of Selected Topics in Signal Processing*, 12(1), 20–34.

Rashedi, E., Nezamabadi-Pour, H., and Saryazdi, S. (2009). GSA: A gravitational search algorithm. *Information Sciences*, 179(13), 2232–2248.

Rojas, R. (2013). *Neural Networks: A Systematic Introduction.* Springer Science & Business Media, Berlin.

Roy, A., Midya, S., Majumder, K., Phadikar, S., and Dasgupta, A. (2017). Optimized secondary user selection for quality of service enhancement of two-tier multi-user cognitive radio network: A game theoretic approach. *Computer Networks*, 123, 1–18.

Russell, S.J. and Norvig, P. (2016). *Artificial Intelligence: A Modern Approach*. Pearson Malaysia, Kuala Lumpur.

Salahdine, F., Kaabouch, N., and El Ghazi, H. (2017a). Techniques for dealing with uncertainty in cognitive radio networks. *IEEE 7th Annual Computing and Communication Workshop and Conference*. 9–11 January 2017, Las Vegas, USA, 1–6.

Salahdine, F., Kaabouch, N., and El Ghazi, H. (2017b). A Bayesian recovery technique with Toeplitz matrix for compressive spectrum sensing in cognitive radio networks. *International Journal of Communication Systems (IJCS)*, 30(15), 1–9.

Saoucha, N.A. and Benmammar, B. (2017). Adapting radio resources in multicarrier cognitive radio using discrete firefly approach. *International Journal of Wireless and Mobile Computing*, 13(1), 39–44.

Shu, Z., Qian, Y., Yang, Y., and Sharif, H. (2016). A game theoretic approach for energy-efficient communications in multi-hop cognitive radio networks. *Wireless Communications and Mobile Computing*, 16(14), 2131–2143.

Supraja, P. and Pitchai, R. (2019). Spectrum prediction in cognitive radio with hybrid optimized neural network. *Mobile Networks and Applications*, 24(2), 357–364.

Surafel, L.T. and Ngnotchouye, J.M.T. (2017). Firefly algorithm for discrete optimization problems: A survey. *KSCE Journal of Civil Engineering*, 21(2), 535–545.

Tang, M. and Xin, Y. (2016). Energy efficient power allocation in cognitive radio network using coevolution chaotic particle swarm optimization. *Computer Networks*, 100, 1–11.

Tounsi, A. and Babes, M. (2017). An efficient joint spectrum and power allocation in cognitive radio networks using a modified firefly algorithm. *International Journal of Communication Networks and Distributed Systems*, 19(2), 214–236.

Tuan, P. and Koo, I. (2016). Throughput maximisation by optimising detection thresholds in full-duplex cognitive radio networks. *IET Communications*, 10(11), 1355–1364.

Tuan, P. and Koo, I. (2017). Robust weighted sum harvested energy maximization for SWIPT cognitive radio networks based on particle swarm optimization. *Sensors*, 17(10), 2275.

Umar, R. and Sheikh, A.U.H. (2012). Cognitive radio oriented wireless networks: Challenges and solutions. *Proceedings of the International Conference on Multimedia Computing and Systems*. 10–12 May 2012. Tangier, Morocco, 992–997.

Wang, B. and Liu, K.J.R. (2011). Advances in cognitive radio networks: A survey. *IEEE J. Selected Topics Signal Process*, 5(1), 5–23.

Wang, D. and Yang, Z. (2016a). A novel spectrum sensing scheme combined with machine learning. *9th International Congress on Image and Signal Processing, BioMedical Engineering and Informatics*. 15–17 October 2016, Datong, China, 1293–1297.

Wang, D. and Yang, Z. (2016b). An advanced scheme with decision tree for the improvement of spectrum sensing efficiency in dynamic network. *9th International Congress on Image and Signal Processing, BioMedical Engineering and Informatics*. 15–17 October 2016, Datong, China, 1288–1292.

Wiering, M. and Van Otterlo, M. (2012). Reinforcement learning. *Adaptation, Learning, and Optimization*, 12, 3.

Woods, W.A. (1986). Important issues in knowledge representation. *Proceedings of the IEEE*, 74(1), 1322–1334.

Wyglinski, A.M., Nekovee, M. and Hou, T. (eds) (2009). *Cognitive Radio Communications and Networks: Principles and Practice*. Academic Press, Cambridge.

Xingjian, L. *et al.* (2018). Intelligent power control for spectrum sharing in cognitive radios: A deep reinforcement learning approach. *IEEE Access*, 6, 25463–25473.

Yang, X.-S. (2005). Engineering optimizations via nature-inspired virtual bee algorithms. International *Work-Conference on the Interplay Between Natural and Artificial Computation*. Las Palmas, Canary Islands, Spain, 317–323.

Yang, X.-S. and Deb, S. (2017). Cuckoo search state-of-the-art and opportunities. *IEEE 4th International Conference on Soft Computing & Machine Intelligence*. 23–24 November 2017, Mauritius, 55–59.

Yin, Z., Wang, Y., and Wu, C. (2017). A multiagent collaborative model for bayesian opportunistic channel accessibility in railway cognitive radio. *International Journal of Performability Engineering (IJPE)*, 13(4), 479–489.

Zadeh, L.A. (1965). Fuzzy sets. *Information and Control*, 8(3), 338–353.

Zaheer, M., Uzma, M., Asif, A., and Qureshi, I. M. (2016). Interference control in cognitive radio using joint beamforming and power optimization by applying artificial bee colony. *19th International Multi-Topic Conference*. 5–6 December 2016. Islamabad, Pakistan, 1–6.

Zhai, L. and Wang. H. (2017). Crowdsensing task assignment based on particle swarm optimization in cognitive radio networks. *Wireless Communications and Mobile Computing (WCMC)*. 4687974, 1–9.

Zhang, M., Diao, M., and Guo, L. (2017). Convolutional neural networks for automatic cognitive radio waveform recognition. *IEEE Access*, 5, 11074–11082.

Zhao, Y. and Morales-Tirado, L. (2012). Cognitive radio technology: Principles and practice. *International Conference on Computing, Networking and Communications*. 30 January–2 February 2012. Maui, USA, 650–654.

Zhu, J., Jiang, D., Ba, S., and Zhang, Y. (2017). A game-theoretic power control mechanism based on hidden Markov model in cognitive wireless sensor network with imperfect information. *Neurocomputing*, 220, 76–83.

10

Cognitive Radio Contribution to Meeting Vehicular Communication Needs of Autonomous Vehicles

Francine KRIEF[1], Hasnaâ ANISS[2], Marion BERBINEAU[2] and Killian LE PAGE[3]

[1] *ENSEIRB-MATMECA, Bordeaux, France*

[2] *IFSTTAR, Bordeaux, France*

[3] *ALTEN, Boulogne-Billancourt, France*

10.1. Introduction

Vehicular communication needs will evolve and experience a boom due to the ascent of connected and autonomous vehicles (AV). Connectivity of various types such as vehicle-to-vehicle (V2V), vehicle-to-infrastructure (V2I), vehicle-to-pedestrian (V2P), and more broadly vehicle to X enables the deployment of a wide variety of applications aiming first of all to improve road safety, and also user comfort during travel. In the context of AVs, new services will emerge, such as platooning, which enables a human-driven car to guide AVs gathered in a convoy through the urban network. Car sharing and the Internet of Things will further broaden the range of services offered. Although currently few, these applications will have various demands in terms of quality of service (QoS) and communications security, which will need to be addressed. Communication link reliability will require self-adaptation of radio access technology, which is made possible by cognitive radio (CR), a technology that is able to detect free frequency bands and adapt its transmission parameters to communication needs and constraints. This concept was introduced by Mitola and Maguire (1999).

CR is defined by its perception, adaptation and cognition capacities. It is characterized by its capacities in terms of spectral interoperability, efficiency, optimization of radio resources and improvement of communications reliability, which are of significant interest to the connected AV. The first prototypes were successfully developed to meet the demands of military communications and those of public security. Besides being able to perceive and adapt to its radio-electrical environment, CR also has learning capacities via artificial intelligence techniques.

Section 10.2 introduces the AV and its main components. Section 10.3 describes applications of vehicle communication and their constraints in terms of QoS. Section 10.4 presents various communication architectures. Section 10.5 highlights the contribution of CR to the vehicular field. This section also presents a state-of-the-art of the main research works in the CR field for vehicular networks. Section 10.6 clarifies our positioning within SERENA project. Finally, section 10.7 concludes this chapter.

10.2. Autonomous vehicles

The AV, the driving of which is partly or fully automated, is currently one of the main technological challenges. Its ascent is expected to radically change the mobility, safety and behavior of road users.

Research on AVs, or rather on automated and connected vehicles, is very active and covers many domains, such as connectivity through digital infrastructure, environment sensing technologies, safe and accurate localization, high definition cartography, the legal field with all regulatory and legal aspects related to responsibility (vehicle owners, manufacturers, State, communities) but also philosophy and ethics. What is the moral responsibility of the developer of artificial intelligence behind the vehicle?

10.2.1. *Automation levels*

The Society of Automotive Engineers (SAE) identifies six levels of driver assistance, ranging from 0 to 5. A vehicle is actually considered autonomous from level 5 up. At lower levels, a vehicle is not yet considered autonomous, as it requires driver presence. The automation levels are described in Table 10.1 (Shladover *et al.* 2014).

10.2.2. *The main components*

A car is referred to as autonomous if its driving is partly or fully automated through various sensors enabling the perception of the environment and connectivity to road infrastructure in order to anticipate the events on the road. The various elements enabling AV to perceive, localize and communicate are summarized in Figure 10.1.

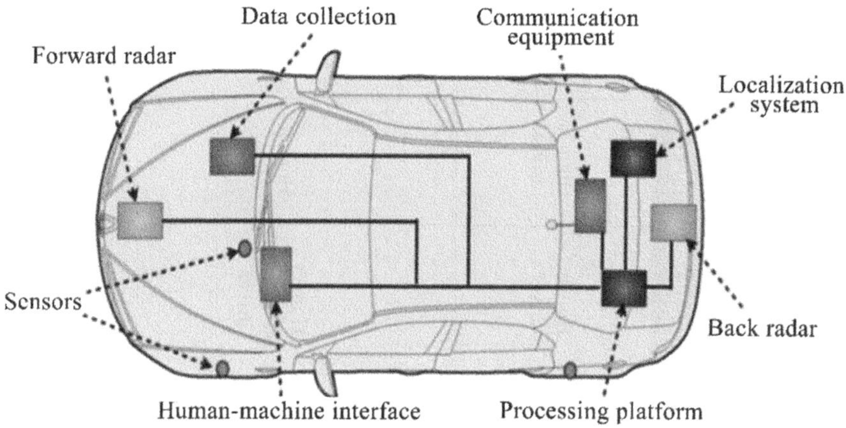

Figure 10.1. *Autonomous vehicle (Hubaux 2005)*

The forward and back perception systems are able to perceive the physical environment in order to prevent a potential collision. These systems are of a multisensor type and include anti-collision radar, lidar and imaging systems in the visible spectrum and sometimes in the infrared spectrum. Certain systems also consider audio sensors. The data collection and processing module is responsible for data recovery. It is directly linked to the control area network (CAN) bus of the vehicle and enables the access to data such as speed, acceleration or temperature and humidity, but also to the state of the vehicle and its environment.

SAE level	Name	Description	Driving execution (steering wheel, acceleration, braking)		Monitoring of driving environment	Driving take-over	System capacity
Human monitoring of the driving environment							
0	No automation	Full time execution by the human driver of all aspects of dynamic driving task, even when reinforced by alert or intervention systems	Human driver	Human driver	Human driver	N/A	
1	Driving assistance	Execution (specific driving mode) by the driving-aid system of maneuvers, either on direction, or on acceleration/deceleration, by means of information on the driving environment. The human driver executes all the other aspects of the dynamic driving task.	Human and system driver	Human driver	Human driver	Certain driving modes	

SAE level	Name	Description	Driving execution (steering wheel, acceleration, braking)		Monitoring of driving environment	Driving take-over	System capacity
2	Partial automation	Execution (specific driving mode) by one or several driving-aid systems of actions both on the direction and on acceleration/deceleration by means of information from the driving environment. The human driver executes all the other aspects of the dynamic driving task.	System	Human driver	Human driver	Certain driving modes	
The automated driving system monitors the driving environment							
3	Conditional automation	Execution (specific driving mode) of all the aspects of the dynamic driving task by an automated driving system with the expectation that the human driver will appropriately respond to an intervention request	System	System	Human driver	Certain driving modes	

SAE level	Name	Description	Driving execution (steering wheel, acceleration, braking)		Monitoring of driving environment	Driving take-over	System capacity
4	Strong automation	Execution (specific driving mode) of all the aspects of the dynamic driving task by an automated driving system, even though the human driver does not appropriately respond to the intervention request	System	System	System	Certain driving modes	
5	Full automation	Full time execution by an automated driving system of all the aspects of the dynamic driving task under all road and environmental conditions that could be managed by a human driver	System	System	System	All driving modes	

Table 10.1. *Various SAE automation levels (source: 2014 SAE International)*

Nowadays, communication equipment relies on the ITS-G5 system and the existing cellular systems. It enables communication between vehicles and infrastructure, but also intervehicle communication. It is particularly responsible for the transmission of alerts to control centers in case of detection of potentially dangerous situations.

The human-machine interface can be used by the driver to interact with the vehicle.

The localization system enables the geo-positioning of the AV, which will be all the more accurate as it is associated with high definition mapping and merging of data issued from multiple sensors.

For further details, the reader is referred to the thematic report "Perspective on the autonomous vehicle" conducted by IFSTTAR (*Institut français des sciences et technologies des transports, de l'aménagement et des réseaux* [French Institute of Science and Technology for Transport, Development and Networks])[1].

10.3. Connected vehicle

Cooperative Intelligent Transport Systems (C-ITSs) based on vehicular communications enable the deployment of new applications. These can be classified into several categories; road safety and entertainment are worth being mentioned here. The needs in terms of QoS and the performances vary depending on the type of application.

10.3.1. *Road safety applications*

Road safety applications are the most critical. It is one of the categories with the highest demands in terms of QoS. With no claim to provide an exhaustive list, examples of applications range from collision detection to accident management, passing through the detection of pedestrians on the road (Dar *et al.* 2010; Cunha *et al.* 2016).

– *Collision detection*: the objective is upstream detection of the risk of collision with an arbitrary obstacle. The most critical points for communication are latency, which must be the lowest possible in order to transmit information as fast as possible, and accuracy of vehicle positioning (mobile or static) and of the obstacle.

1 Available at : https://www.ifsttar.fr/ressources-en-ligne/espace-science-et-societe/mobilites/dossiers-thematiques/vehicule-autonome/.

– *Accident management*: the objective is to detect accidents that are ongoing or predictable in the very short term in order to change driver behavior, for example by suggesting a new route that bypasses the accident. In this case, a sufficiently wide radio range is required for the best possible anticipation. Latency must also be very low to enable making a decision as rapidly as possible. Finally, referring to throughput, it must be sufficiently significant in order to transmit all the required information, following the encountered accident.

– *Detection of pedestrians crossing the road*: the objective is to notify the driver on the presence of a pedestrian unexpectedly arriving on the road. The driver (or the AV) can thus brake or change the direction of its vehicle to avoid any accident with the pedestrian. Latency must therefore be sufficiently low to make a real-time reaction possible and thus avoid an accident. A range of at least 200 m is required to best anticipate the passage of a pedestrian.

The referenced works of Dar *et al.* (2010), Amjad *et al.* (2018), MacHardy (2018), and Mir and Filali (2018) propose certain performance characteristics required for the implementation of these use cases, as summarized in Table 10.2.

Application	Latency (ms)	Throughput (Mo/s)	Distance (m)
Collision detection	10–100	From 0.0625 to 87.5	Urban environment: 500 Highway: 2,000
Accident management	Of the order of millisecond	From 0.125 to 0.75	From less than 500 to approximately 1,000
Pedestrian detection	20–100	From 0.0625 to 87.5	≥200

Table 10.2. *Constraints expressed for road safety vehicular applications*

10.3.2. *Entertainment applications*

Entertainment applications are the least critical in terms of QoS. By their own nature, these applications do not require very high priorities, unlike the previously mentioned applications. The needs are considered on a case by case basis. Let us consider two different examples from Campolo *et al.* (2017): video streaming and Internet browsing.

– *Video streaming*: this application broadcasts or reads video flows. Latency must be the lowest possible to provide the user with good quality experience. Throughput must be dimensioned according to the expected quality. For very good video quality, throughput must be significant, while it can be smaller for lower video quality. The radio range must be wide enough to enable access to content during a long route without any interruption.

– *Internet browsing*: the user must have access to Internet pages. Latency is less important compared to other applications, while being low enough to meet the quality of user experience. The throughput required for smooth browsing may depend on the visited content, but the latter does not need to be as significant as for other applications. Finally, the radio range must be wide to enable continuous browsing.

The works of Campolo *et al.* (2017) and Dar *et al.* (2010) estimate the minimal required performances, as summarized in Table 10.3.

Application	Latency (ms)	Throughput (Mo/s)	Range (m)
Video streaming	Lowest possible	1.875 (for UHD video)	>1,000
Internet browsing	100 ms	0.0625	>1,000

Table 10.3. *Constraints expressed for vehicular entertainment applications*

10.4. Communication architectures

C-ITS vehicular networks rely on a set of ISO (International Organization for Standardization) and ETSI (European Telecommunication Standards Institute) standards defining the architecture of each component (vehicle, roadside unit, road management center, etc., see Figure 10.2) on the basis of an ITS (Intelligent Transport System) protocol stack and an OSI (Open Systems Interconnexion) protocol stack.

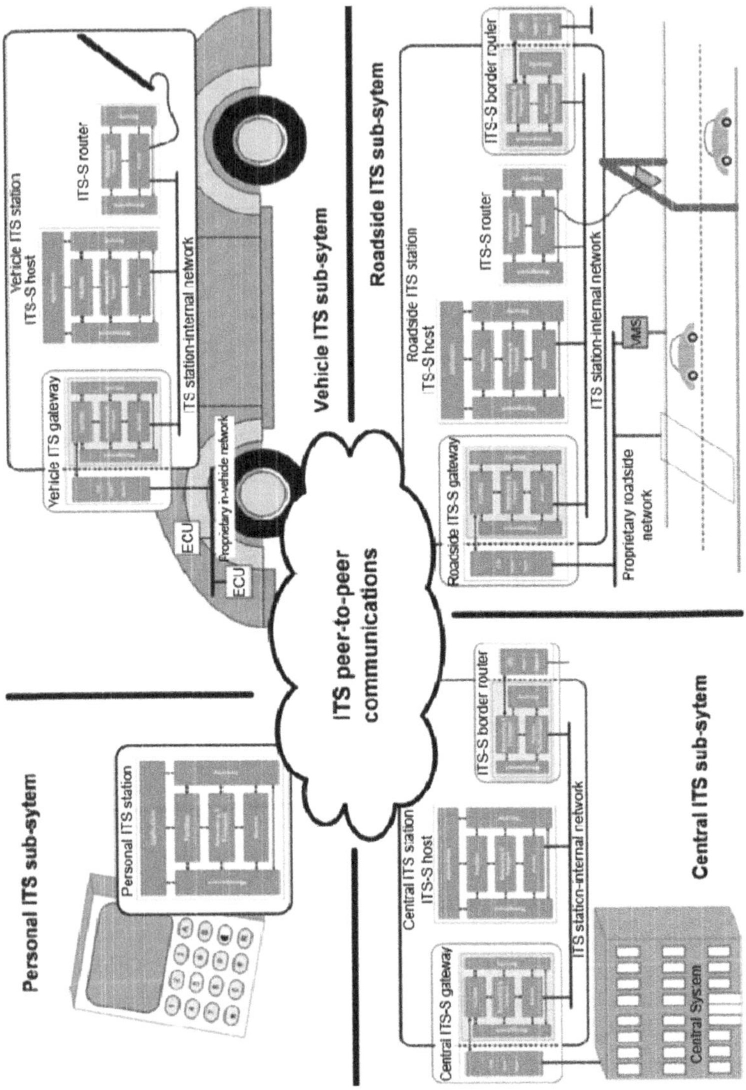

Figure 10.2. *Illustration of C-ITS systems (ETSI 2010)*

Each subsystem illustrated in Figure 10.2 is defined from the same architecture, which is composed of (see Figure 10.3):

– an "Access" layer, which represents OSI layers 1 and 2;

– a "Network and transport" layer, which represents OSI layers 3 and 4;

– an "Installation" layer, which represents OSI layers 5, 6 and 7;

– an "Application" layer, which manages the production of C-ITS service;

– a cross "Security" layer for message securing;

– a cross "Management" layer, which manages the interaction between various layers.

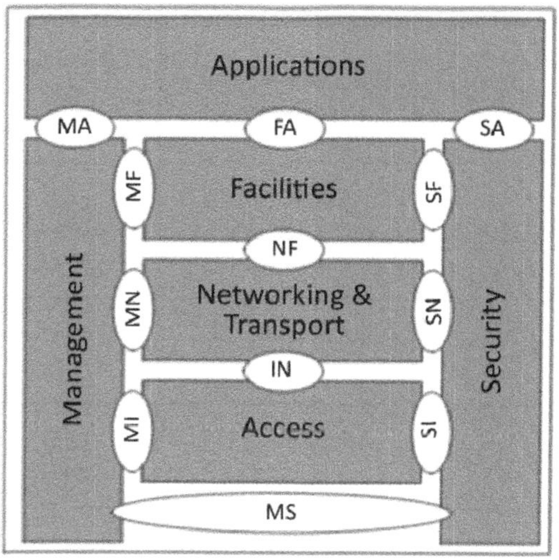

Figure 10.3. *Reference architecture of an ITS station (ETSI 2010)*

In terms of access, two types of communication are used:

– in *ad hoc* mode, ITS-G5; in this case, the transport is made either in IP or more generally using geo-networking (ETSI 2014);

– cellular (3G/4G); the link is then exclusively IP.

These two access modes are often coupled in order to increase service coverage; this is the hybrid mode.

10.4.1. *ITS-G5*

The ITS-G5 system is governed by a set of ETSI standards based, for the physical layer/support access control, on the IEEE 802.11p (IEEE 2010) standard.

ETSI allocated three frequency bands in the band of 5 GHz for C-ITS systems. Each band is divided into 10 MHz channels. The first band, referred to as ITS-G5A, of 30 MHz, is dedicated to road safety applications. The second band of 20 MHz, referred to as ITS-G5B, is dedicated to other applications. The last band of 20 MHz (ITS-G5D) is dedicated to future use.

	Frequency interval	Use	Regulations	Harmonized standard
ITS-G5A	5,875–5,905	ITS applications for road safety	Decision of the commission	EN 302571
ITS-G5B	5,855–5,875	ITS applications not related to road safety	ECC Recommen-dation	EN 302571
ITS-G5C	5,470–5,725	RLAN (BRAN, WLAN)	ERC decision	EN 301893
ITS-G5D	5,905–5,925	Applications ITS futures	ECC decision	EN 302571

Table 10.4. *ITS-G5 system*

Channel type	Central frequency (MHz)	Channel number IEEE 802.11	Channel spacing (MHz)	Throughput by default (Mbit/s)	Limit power rate (dBm PIRE)	Limit power density rate (dBm/MHz)
G5-CCH	5,900	180	10	6	33	23
G5-SCH1	5,880	176	10	6	33	23
G5-SCH2	5,890	178	10	12	23	13

<table>
<tr><th>Channel type</th><th>Central frequency (MHz)</th><th>Channel number IEEE 802.11</th><th>Channel spacing (MHz)</th><th>Throughput by default (Mbit/s)</th><th>Limit power rate (dBm PIRE)</th><th>Limit power density rate (dBm/MHz)</th></tr>
<tr><td>G5-SCH3</td><td>5,870</td><td>174</td><td>10</td><td>6</td><td>23</td><td>13</td></tr>
<tr><td>G5-SCH4</td><td>5,860</td><td>172</td><td>10</td><td>6</td><td>0</td><td>– 10</td></tr>
<tr><td>G5-SCH5</td><td>5,850</td><td>182</td><td>10</td><td>6</td><td>0</td><td>– 10</td></tr>
<tr><td>G5-SCH6</td><td>5,910</td><td>184</td><td>10</td><td>6</td><td>0</td><td>– 10</td></tr>
<tr><td rowspan="2">G5-SCH7</td><td rowspan="2">As described in the band: 5,470 MHz to 5,725 MHz</td><td rowspan="2">94–145</td><td rowspan="2">Several</td><td rowspan="2">Depends on channel spacing</td><td>30 (DFS master)</td><td>17</td></tr>
<tr><td>23 (DFS slave)</td><td>10</td></tr>
</table>

Table 10.5. *Table of ITS-G5 frequency allocation in Europe (IEEE 2010)*

Taking over numerical throughputs of 3, 6 and 12 Mbps is compulsory for ITS stations.

Support access control (MAC)

To enable the use of 802.11 standard in vehicular communications, the following functionalities were modified:

– authentication of the MAC sublayer and association processes are deactivated;

– energy saving is not authorized;

– the security of 802.11 is not supported.

10.4.2. *LTE-V2X*

In 2016, the 3GPP standardization group published specifications for V2X communications relying on long-term evolution (LTE) technology referred to as "LTE V2X" (or C-V2X) to distinguish them from V2X 802.11p technology. Besides direct communication (V2V, V2I), C-V2X also supports long-distance communication on a cellular network (V2N). LTE-V2X standard supports two radio interfaces:

– cellular interface Uu;

– PC5 interface in charge with V2V communications. In mode 3, the cellular network allocates and manages the radio resources used by vehicles for their communications. On the other hand, in mode 4, the vehicles autonomously select the radio resources.

At the end of 2018, 3GPP published release 15: the first specification for 5G NR (New Radio). 5G follows LTE with a period of transition in which LTE will be upgraded because of 5G. One of the radio differences compared to LTE is a larger frequency band (from 700 MHz to 70 GHz), segmented according to the types of applications or environments (dense urban to rural). Moreover, the information will be processed as close as possible to the final user in order to improve system performances. Latencies are expected to be below 1 ms.

10.4.3. *Hybrid communication*

In Mannoni *et al.* (2019), after simulation of various communication modes (LTE-C-V2X, ITS-G5) the authors show that, despite a more significant range and an often higher throughput for the cellular mode, the performances of ITS-G5 remain higher than those of applications always requiring low latency, namely road safety applications. Nevertheless, in order to increase the penetration rate of C-ITS services, it may be relevant to use short- and long-range communications simultaneously for entertainment or information applications. Short-range communications remain dedicated to services related to transport security. The major hybridization principle is to enable the receipt, whatever the network used, of the same information. C-ITS messages are the same, irrespective of the chosen communication medium, and pass through a central unit enabling a geographical connection between all the cellular units. This requires that all the higher layers of the C-ITS stack are identical (application, installation and security layers). There are variations at the level of the network and transport layer and the access layer.

10.5. Contribution of CR to vehicular networks

CR is the combination of a so-called “software” radio with a decision-making module. This combination enables the dynamic adaptation of the radio system to its electromagnetic environment. As already mentioned, the concept of CR was introduced by J. Mitola in 1999 (Mitola and Maguire 1999). CR can be used in the vehicular domain in order to always remain connected despite the mobility and the variation of the type of telecommunication network available at a given moment along the vehicle path. This section first describes the specificities of CR, then presents its contribution to vehicular networks. This is followed by a state-of-the-art of the main research works in the field of CR for vehicular networks.

10.5.1. *Cognitive radio*

The main contribution of CR is better management of the radio-electrical spectrum, using frequency bands that are left vacant at a given instant, because of dynamic access to the spectrum. For this purpose, CR generally distinguishes between two user categories. The users referred to as *primary* can use the licensed bands (frequency bands of mobile operators, etc.) at any moment, based on adequate subscription. The second category of users is that of *secondary* users. In this category, the users cannot use the licensed bands unless they are free at the moment of use, there is no risk of generating interference and it is possible to change the band as soon as a primary user shows up. To be able to dynamically access the spectrum, the CR system must make its decisions based on the policy of spectrum allocation defined by the regulatory authorities. In general terms, CR can adapt to its environment because of electromagnetic environment sensing modules, spectrum analysis, wave shape recognition, decision making and the capacity for dynamic reconfiguration of the parameters of the radio system (throughput, frequencies, modulation, coding, etc.). The operation of CR follows a cycle known as a "cognition loop", as illustrated in Figure 10.4.

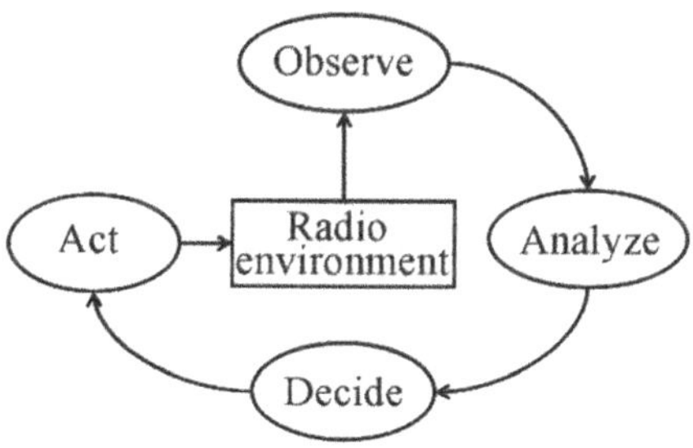

Figure 10.4. *Cognition loop*

CR is a very active research domain. Indeed, the capacity of a radio system to analyze its environment, to choose the bands in which it can emit and to reconfigure are increasingly important characteristics for the cohabitation of various wireless telecommunication systems, for the optimization of services depending on mobility and on network load, for information routing through various networks, etc.

For further details on CR systems, the interested reader is invited to refer to Arslan (2007), Doyle (2009) and Palicot *et al.* (2011).

In the context of 5G development, the 5GPPP group, financed by the European Commission, has played a very important role in prestandardization since 2015. In this context, dynamic access to various frequency bands and to various radio access technologies is possible (extended Dynamic Spectrum Access [eDSA], MAC framework). The architecture of this new multitechnology access protocol is under development and relies on LTE-A. The centralized radio resources manager (cRRM), the spectrum manager and the operation and administration and management (OAM) play a fundamental role, which could be related to the cognitive manager that will be introduced further on (5G PPP Architecture Working Group 2017).

10.5.2. *CR-VANET*

CR-VANET, which stands for Cognitive Radio for Vehicular Ad-hoc Network, or CRAVENET, which stands for Cognitive Radio Assisted Vehicular Network, is an evolution of the *ad hoc* vehicular network.

This type of network uses the capacities of CR to enable a better management of the radio spectrum, guaranteed connectivity and improvement in the available bandwidth and QoS.

Singh *et al.* (2014) and Eze *et al.* (2017) propose an overview of the research works on CR-VANET. It is a very active domain. In Singh *et al.* (2014), the authors propose a taxonomy of the main problems studied in the literature. In Eze *et al.* (2017), work in the routing field, MAC layer and security completes the taxonomy. The work related to simulators and to evaluation platforms is also presented.

Figure 10.5 illustrates the main fields of research dealt with.

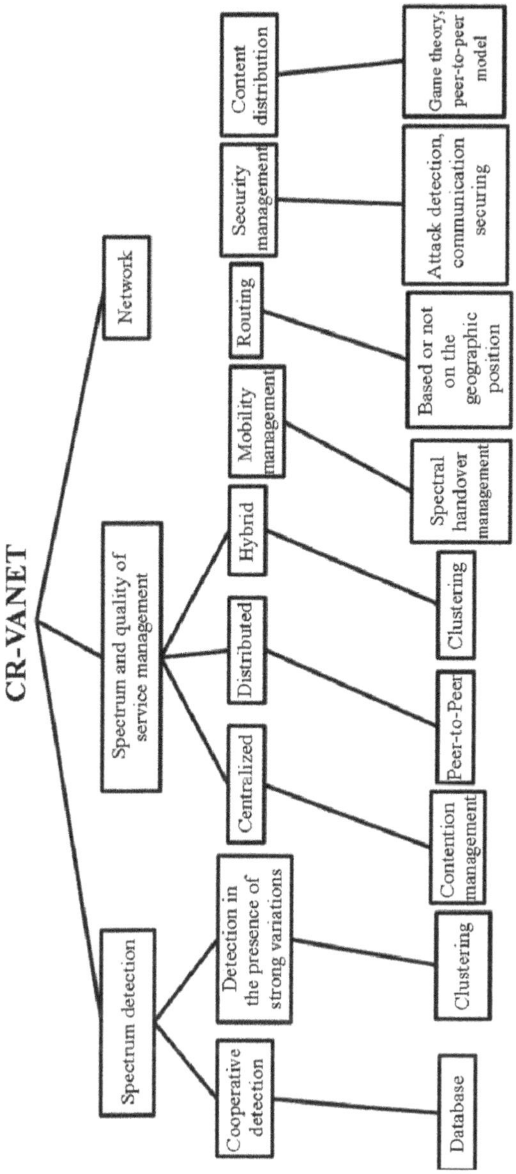

Figure 10.5. *The main problems related to CR in the literature (Singh et al. 2014; Eze et al. 2017)*

The following sections enumerate the main publications in the fields indicated in Figure 10.5, namely those on detection of spectrum sensing, identification of the type of system present in the analyzed band, spectrum management, QoS and network-related work.

10.5.2.1. *Detection of spectrum sensing*

The main theme of spectrum detection, also referred to as "spectral survey", is to enable the best possible detection of spectrum sensing by primary users, taking into account the vehicular characteristics. There are many such works. Generally speaking, there are several approaches, such as narrow band or wide band, cooperative or non-cooperative approaches and blind or non-blind approaches. When a user is detected in the surveyed band, it is important to detect the type of modulation in order to be able to identify the network to which the detected primary user belongs.

The field of spectral survey and modulation detection has been very active in recent years. Even a standard was proposed, namely the IEEE 802.22 (IEEE 2019) standard. A certain number of publications focused on radio environment detection when the activity of channels strongly varies in time, particularly in the railway sector. Blind spectral survey methods were developed by Hassan *et al.* (2014), notably in the presence of impulsive noise, and Bouallegue *et al.* (2018). Various modulation recognition methods were proposed in Hassan *et al.* (2010, 2012), Kharbech *et al.* (2013) and Kharbech (2018). This theme is very important, since a mobile environment composed of vehicles leads to frequent variations of spectrum sensing, particularly depending on the environment (highway, country road, etc.).

The cooperative approach relies on the use of a database or clustering, as proposed in Li *et al.* (2018). The results of the detection made by several vehicles is combined and analyzed in view of the best allocation of frequencies among vehicles. This approach is all the more interesting as the available frequency bands may vary according to the geographical position of the vehicle. Data can be merged in a distributed manner among vehicles or in a centralized manner, passing through an infrastructure, such as the roadside units.

10.5.2.2. *Spectrum and QoS management*

In this category, research focuses on the methods enabling the best management of the spectrum, while achieving a certain QoS. The proposed methods are centralized, distributed or hybrid.

The centralized method involves the transmission of various spectrum and QoS-related information to an infrastructure in charge of collecting and analyzing information in order to decide on the vehicle behavior.

The distributed method involves relying on its pairs and on itself for the best spectrum and QoS management. An example dealing with peer-to-peer and therefore distributed distribution is presented in Bradai *et al.* (2014). Video content is broadcast between vehicles by choosing the best channel available at an instant t, by conducting a dynamic analysis of the quality of channels and relying on a peer-to-peer network for data transmission. Channel selection relies first of all on the best available dedicated short-range communications (DSRC) channel. Otherwise, CR supports the choice of the best channel among available channels that do not belong to DSRC. The selection criterion is the quality of the received signal strength indication (RSSI) connection.

With regard to the hybrid method, which is used in Niyato *et al.* (2011), this relies on a structure such as road infrastructure and on the vehicles forming an opportunistic network to best manage the spectrum and the QoS. In Niyato *et al.* (2011), the authors try to reduce the infrastructure load by relying on the cluster leaders for the communication between clusters and with infrastructure in order to make the best decisions.

10.5.2.3. *Network*

All the classical network themes are applied here to vehicular networks, namely mobility management, routing, content distribution and security management.

In the context of mobility management, handover is taken into account to maintain an active connection and prevent channel interferences. Hence, in Kumar *et al.* (2017), the authors propose a solution for spectral handover management.

Several methods are proposed for message routing: based on geographical position, without knowledge on geographical position, etc. Hence, in Usha and Ramakrishnan (2019), the authors improve the MPR OLSR algorithm by taking into account inactive channels in order to use them while choosing relay nodes with the highest number of new neighbors.

The work on content distribution focuses on the use of game theory or the peer-to-peer approach. Hence, in Tian *et al.* (2019), the authors use an evolutionary game approach that assigns licensed channels to secondary users based on price evolving in time and load balancing using this channel. Concerning the use of the peer-to-peer model, the work of Bradai *et al.* (2014) is worth mentioning, which relies on the notion of a neighborhood to broadcast video fragments and thus enable video content broadcast through the network.

Finally, network security and the protection of user's private life are very important research subjects. Indeed, road safety messages are essential for the operation of a CR-VANET network to the same extent as in a VANET network (Mitra *et al.* 2016; Wei *et al.* 2016). It is therefore important that these messages (accident management, collision management, etc.) reach their destination. In Mitra *et al.* (2016), the focus is on the detection of black hole attacks in the network and their elimination. These attacks involve the creation of a fictitious virtual node through which all communications enter but do not exit in order to disrupt communications between vehicles. The proposed solution is the detection of these attacks and, once the black hole is identified, an alternative path is set up, bypassing the black hole and enabling communications to resume. In Wei *et al.* (2016), the authors proposed the use of a light cloud in association with a roadside unit infrastructure (UBR) for securing communications in a CR-VANET. Communications are also secured by using a new service, referred to as "Spectrum Sensing as a Service" (SaaS), enabling a detection of the cooperative spectrum using the implemented cloud.

The following presents research that we intend to conduct in the field of CR for vehicular networks.

10.6. SERENA project: self-adaptive selection of radio access technologies using CR

SERENA is a joint research project conducted by LaBRI (*Laboratoire bordelais de recherche en informatique* [Bordeaux Laboratory for Computer Science Research]) and IFSTTAR (*Institut français des sciences et technologies des transports, de l'aménagement et des réseaux* [French Institute of Science and Technology for Transport, Development and Networks]). It also benefits from the support of the SysNum cluster (Numerical systems) of Idex Bordeaux (*Initiative d'excellence de l'université de Bordeaux* [Initiative of Excellence of the University of Bordeaux]). The general objective of this project, starting at the end of 2019, is the improvement of vehicular communications to enable the deployment of new services, in particular those required for the autonomous vehicle. This involves, in particular, the definition of a self-adaptive selection mechanism of radio access technologies in order to always be best connected.

The proposed self-adaptation mechanism relies on the specification of new decision-making algorithms that are capable of taking into account various constraints, such as QoS, security, energy consumption or user preferences (such as the cost of the communication, for example). The prediction of the probable evolution of the service level will also enable proactive actions in order to provide service continuity in mobility.

The SERENA project enables the optimization and improvement of the quality of data exchanges between vehicles and with the infrastructure. Moreover, because of the retained approach, the evolutions of radio access technologies can be taken charge of due to the use of CR and the virtualization of network functions.

10.6.1. *Presentation and positioning*

The SERENA project proposes a solution that should enable the connected vehicle to autonomously select in real time the radio access technologies that best meet the needs of each type of communication flow (V2I and/or V2V) and for quite precise use cases. For this purpose, the devised solution relies on the concept of CR and on the introduction of an appropriate cognitive manager located in the embedded terminal. This solution will first be validated by simulations, considering several use cases related to the AV, then at real size, considering software radio maps.

Among the research on the dynamic selection of the best access network for vehicular communications, the work of Singh *et al.* (2014) and Kumar *et al.* (2017) deal essentially with V2V communications using CR. Within the ANR PLATA (*Plateforme télématique multistandard programmable pour l'automobile* [Programmable Telematics On-board Multistandard Radio Platform]) project, the use of embedded software-defined radio (SDR) systems for V2V and V2I communications was considered (Haziza *et al.* 2013).

In the railway sector, a solution on the middleware-based applicative layer for choosing the best wireless network available for V2I communications depending on various criteria was proposed (Billion *et al.* 2008). This concept was also proposed in Amanna *et al.* (2010). In the CORRIDOR (COgnitive Radio for RaIlway through Dynamic and Opportunistic spectrum Reuse) project, works were conducted for the self-adaptation of radio access technology for V2I communication applications for high-speed trains (Berbineau *et al.* 2014).

To our knowledge, as of today, there are no works in the road sector related to the self-adaptation of radio access technology to the needs of V2X (V2I and V2V) communications using CR intelligence and the virtualization of network functions.

10.6.2. *General architecture being considered*

The general architecture of the proposed system relies on the use of CR and the virtualization of network functions.

10.6.2.1. *Cognition loop and cognitive manager*

As previously mentioned, CR is a technology well adapted to the self-adaptation of radio access technology to meet vehicular communication needs.

The objective here is to adapt the operation of the communication module so that it can, on the one hand, select the best access technology/technologies in the presence of constraints and use cases and, on the other hand, consequently adapt its communication parameters. For this purpose, a cognition loop and an appropriate cognitive manager need to be specified.

Specification of the cognition loop

Using the knowledge resulting from the stages of electromagnetic environment analysis, CR makes reconfiguration decisions in a dynamic manner, depending on certain predefined objectives, so that the efficiency of spectrum use is improved with no human intervention (cognition loop).

Within the SERENA project, decision making may lead, for example, to stopping certain low priority flows in order to protect the more critical flows or to a modification of the level of communications security. It is therefore important to specify the cognition loop so that it can display the theme of the SERENA project, as illustrated in Figure 10.6.

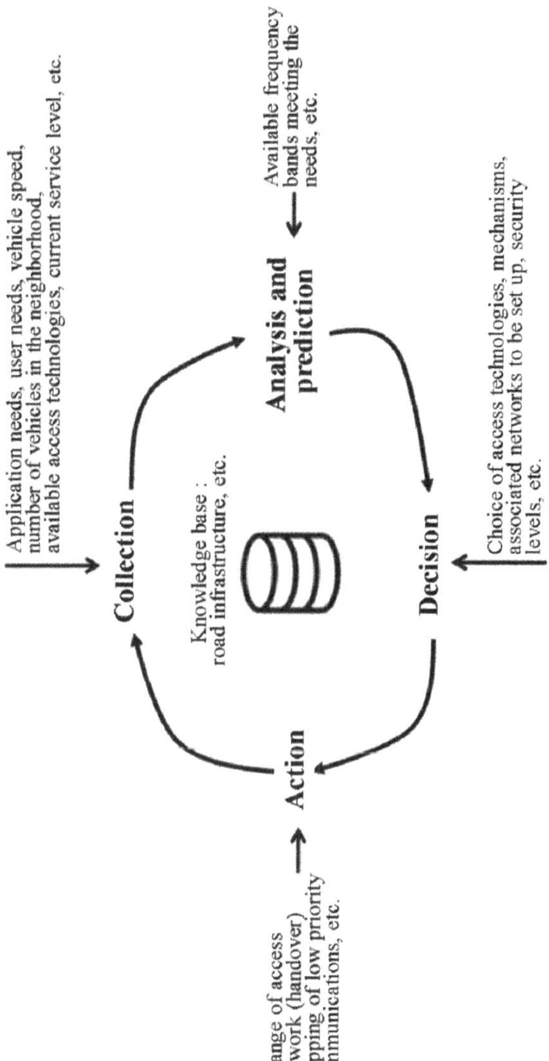

Figure 10.6. *Example of cognitive loop of the SERENA project*

This cognition loop, as already indicated, includes four large stages:

– *collection (observation)*: in this phase, data useful to decision making are collected. Examples of useful data include speed, position of the vehicle and radio environment (cellular networks present in the environment). Static data such as QoS constraints associated with a type of application are previously stored in the knowledge base;

– *analysis and prediction*: an analysis of gathered data is conducted using the knowledge base. Machine learning techniques can be used to predict, for example, the availability time of various frequency bands. An initial filtering can also be done considering only the frequency bands that are expected to meet the communication needs;

– *decision*: this stage enables the selection of the radio technology and communication protocols to be used, or the level of security to be applied relying on data resulted from the previous stage. The decision needs to take into account the short term forecast of the availability of connections between two vehicles or the availability of the telecommunications infrastructure in order to guarantee service continuity. Considerations concerning the near future may rely on information on vehicle motion, on the motions of neighboring vehicles or on infrastructure knowledge (information contained in the knowledge base);

– *action*: this stage enables a set of actions such as radio reconfiguring (change of frequency band, possible modification of the transmission parameters, etc.) according to the characteristics of the access technology retained during the previous stage, stopping low priority communications to benefit from more bandwidth, increasing the security level, etc.

Cognitive manager specification

A cognitive manager uses various cognitive engines (Ouattara 2014) to set up a cognition loop. These engines are in charge of measurement, analysis, reasoning, learning, decision making and adaptation. The content of each of the various engines required for the set-up of the cognition loop of the SERENA project needs to be specified. It is also important to specify the content of the knowledge base. This contains, on the one hand, static information such as the location of roadside units or the networks present in the area and, on the other hand, calculated or forecast information from the cognitive engines.

10.6.2.2. *Network function virtualization*

Network function virtualization (NFV) is nowadays a major evolution of networks. In combination with the software-defined networking (SDN) technique, it offers many advantages to network operators, particularly rapid deployment, higher flexibility and better adaptation to context. The introduction of these two technologies is deeply transforming networks and will contribute to accelerating the introduction of the autonomous car (Mendiboure *et al.* 2019). In particular, 5G relies on SDN/NFV to implement network slicing, which is a logical slicing of the network, to make possible the taking charge of various categories of services such as ITS services with low latency and high availability/reliability. There are still many challenges, such as the orchestration of slices, the management of network functions and the security of SDN networks (Foukas *et al.* 2017).

The service life of a vehicle is relatively long (several years) in relation to the current evolution of radio technologies. Hence, many new communication technologies may emerge within this period of time. The NFV approach is an important asset for the SERENA project, as it ensures the continued existence of the retained solution in relation to technological evolutions, particularly of new radio access means. Indeed, the virtualization of the vehicle communication module will make it possible to readily add new communication functionalities through the addition of software functions. This approach is all the more facilitated by the fact that radio communications rely within this project on SDR.

10.6.3. *The main stakes*

The SERENA project raises several challenges, such as the real-time identification and selection of the access technology in a highly changing environment (variation of the speed of the vehicle, connectivity density, number of available access networks, etc.), the capacity to support several simultaneous access technologies for the same communication (to secure the radio link in case of AV, for example), real-time monitoring of the level of service offered and proactive decision making to maintain the required service level.

The main risk incurred is removing excessively low performances that do not meet the needs of strong mobility. Nevertheless, the maximum authorized speed of the autonomous car is at present limited to several dozen kilometers per hour.

The second risk concerns the current difficulties encountered by the integration of CR with the retained access technologies. This risk may be eliminated with the deployment of NFV (OPNFV)[2] solutions and the virtualization of access protocols in particular (Riggio *et al.* 2016).

10.7. Conclusion

AV is being developed and new applications are emerging. These applications have specific communication needs, either in terms of latency, range or bandwidth. For the best operation of these applications, it is important to select the radio access technology that is best adapted to their needs and therefore to study the advantages and drawbacks of the architectures on which these access technologies rely.

The radio spectrum being poorly used, the use of CR makes it possible to find a solution to the needs of communication continuity in the vehicular environment. This chapter illustrated the main contributions of CR to vehicular domain, such as the improvement of network coverage or the improvement in available bandwidth.

In a CR-VANET, CR can bring its own stakes. For example, in Singh *et al.* (2014), the stakes may relate to the reliability of urgent message communication, the management of dynamic topology, the scarcity of available bands, the management of new communication technologies and the adaptation to various environments (highway, urban area, rural area, etc.) and spectrum distribution.

Within the SERENA project, CR associated with virtualization (NFV) and reconfiguration (SDN) of network functions will enable the vehicle communication module to select the best radio access technology, adapt its radio parameters and be resilient in front of technological evolutions.

The use of AI tools as decision-aiding tools goes hand-in-hand with the development of the AV and machine learning algorithms, in particular, will attract significant enthusiasm in this field.

10.8. References

5G PPP Architecture Working Group (2017). View of 5G architecture (version 2.0) [Online]. Available at: https://5g-ppp.eu/wp-content/uploads/2017/07/5G-PPP-5G-Architecture-White-Paper-2-Summer-2017_For-Public-Consultation.pdf.

Amanna, A., Gadniok, M., Price, M.J., Reed, J.H., Siriwongpairat, W.P., and Himsoon, T.K. (2010). Railway cognitive radio. *IEEE Vehicular Technology Magazine*, 5(3), 82–89.

2 Open Platform for Network Function Virtualization, available at: https://www. opnfv.org/.

Amjad, Z., Sikora, A., Hilt, B., and Lauffenburger, J.-P. (2018). Low latency V2X applications and network requirements performance evaluation. *IEEE Intelligent Vehicles Symposium (IV)*. IEEE, Changsu.

Arslan, H. (2007). *Cognitive Radio, Software Defined Radio, and Adaptative Wireless Systems*. Springer, Dordrecht.

Berbineau, M, Masson, E, Cocheril, Y, Kalakech, A, Ghys, JP, Dayoub, I, Kharbech, S, Zwingelstein-Colin, M, Simon, E, Haziza, N, Painchault, R, Gautier, R, Hassan, K, De La Rosa, KJ, Radoi, E, Nussbaum, D, Knopp, R, Bonnin, JM, Singh, KD, Lee, JH, Philippe, H, Ghannoum, H, Sanz, D. and Massy, P. (2014). Cognitive radio for high speed railway through dynamic and opportunistic spectrum reuse. *Transport Research Arena (TRA) 5th Conference: Transport Solutions from Research to Deployment*. 14–17 Apr 2014. TRA, Paris.

Billion, J., Van den Abeele, D., Gransart, C., and Berbineau, M. (2008). ICOM: Toward integrated communications for global railway systems. *World Congress on Railway Research*. WCRR, Seoul.

Bouallegue, K., Dayoub, I., Gharbi, M., and Hassan, K. (2018). Blind spectrum sensing using extreme eigenvalues for cognitive radio networks. *IEEE Communications Letters*, 22(7), 1386–1389.

Bradai, A., Ahmed, T., and Benslimane, A. (2014). ViCoV: Efficient video streaming for cognitive radio VANET. *Vehicular Communications*, 1(3), 105–122.

Campolo, C., Molinaro, A., Iera, A., and Menichella, F. (2017). 5G network slicing for vehicle-to-everything services. *IEEE Wireless Communications*. 24(6), 38–45.

Cunha, F., Villas, L., Boukerche, A., Maia, G., Viana, A., Mini, R.A., and Loureiro, A.A. (2016). Data communication in VANETs: Survey, applications and challenges. *Ad Hoc Networks*, 44, 90–103.

Dar, K., Bakhouya, M., Gaber, J., and Wack, M. (2010). Wireless communication technologies for ITS applications. *IEEE Communications Magazine*, May, 156–162.

Doyle, L. (2009). *Essentials of Cognitive Radio*. Cambridge University Press, Cambridge.

ETSI (2010). Intelligent Transport Systems (ITS); communications architecture. Document, ETSI, Sophia Antipolis.

ETSI (2014). Vehicular communications; GeoNetworking; Part 4: Geographical addressing and forwarding for point-to-point and point-to-multipoint communications; Sub-part 1: Media-independent functionality. Document, ETSI, Sophia Antipolis.

Eze, J., Zhang, S., Liu, E., and Eze, E. (2017). Cognitive radio technology assisted Vehicular AdHoc Networks (VANETs): Current status challenges, and research trends. *23rd International Conference on Automation and Computing*. ICAC, Huddersfield.

Foukas, X., Elmokashfi, A., Patounas, G., and Marina, M.K. (2017). Network slicing in 5G: Survey and challenges. *IEEE Communications Magazine*. IEEE, 55(5), 94–100.

Hassan, K., Dayoub, I., Hamouda, W., and Berbineau, M. (2010). Automatic modulation recognition using wavelet transform and neural networks in wireless system. *EURASIP Journal on Advances in Signal Processing*. IEEE. Article number: 532898 (2010), 1–13.

Hassan, K., Dayoub, I, Hamouda, W., Nzeza, C.N., and Berbineau, M. (2012). Blind digital modulation identification for spatially-correlated MIMO systems. *IEEE Transactions on Wireless Communications*, 11, 683–693.

Hassan, K., Gautier, R., Dayoub, I., Berbineau, M., and Radoi, E. (2014). Multiple-antenna-based blind spectrum sensing in the presence of impulsive noise. *IEEE Transactions on Vehicular Technology*, 63(5), 2248–2257.

Haziza, N., Kassab, M., Knopp, R., Harri, J. and Kaltenberger, F. *et al.* (2013). Multi-technology vehicular cooperative system based on software defined radio (SDR). *Fifth Workshop on Communication Technologies for Vehicles*. Nets4cars, Vilnius.

Hubaux, J.-P. (2005). Vehicular networks: How to secure them. MiNEMA Summer School. MiNEMA, Klagenfurt.

IEEE (2010). 802.11p. Local and metropolitan area networks – Specific requirements – Part 11: Wireless LAN medium access control (MAC) and physical layer (PHY) specifications amendment 6: Wireless access in vehicular environments. Document, IEEE.

IEEE (2019). IEEE 802.22. Draft standard for information technology – Local and metropolitan area networks – Specific requirements – Part 22: Cognitive radio wireless regional area networks (WRAN) medium access control (MAC) and physical layer (PHY) specifications: Policies and procedures for operation in the bands that allow spectrum sharing where the communications devices may opportunistically operate in the spectrum of the primary service. Document, IEEE.

Kharbech, S., Dayoub, I., Simon, E., and Zwingelstein-Colin, M. (2013). Blind digital modulation detector for MIMO systems over high-speed railway channels. *Communication Technologies for Vehicles*, 232–241.

Kharbech, S., Dayoub, I., Zwingelstein-Colin, M., and Simon, E.P. (2018). Blind digital modulation identification for MIMO systems in railway environments with high-speed channels and impulsive noise. *IEEE Transactions on Vehicular Technology*, 67(8), 7370–7379.

Kumar, K., Prakash, A., and Tripathi, R. (2017). A spectrum handoff scheme for optimal network selection in NEMO based cognitive radio vehicular networks. *Wireless Communications and Mobile Computing*, 2017.

Li, X., Song, T., Zhang, Y., Chen, G., and Hu, J. (2018). A hybrid cooperative spectrum sensing scheme based on spatial-temporal correlation for CR-VANET. *IEEE 87th Vehicular Technology Conference (VTC Spring)*. IEEE, Porto.

MacHardy, Z., Khan, A., Obana, K., and Iwashina, S. (2018). V2X access technologies: Regulation, research, and remaining challenges. *IEEE Communications Surveys & Tutorials*, 20(3), 1858–1877.

Mannoni, V., Berg, V., Sesia, S., and Perraud, E. (2019). A comparison of the V2X communication systems: ITS-G5 and C-V2X. *IEEE Vehicular Technology Conference (VTC Spring)*. IEEE, Kuala-Lumpur.

Mendiboure, L., Chalouf, M.A., and Krief, F. (2019). Edge computing based applications in vehicular environments: Comparative study and main issues. *Journal of Computer Science and Technology*, 34(4), 869–886.

Mir, Z.H. and Filali, F. (2018). Applications, requirements, and design guidelines for multi-tiered vehicular network architecture. *10th Wireless Days Conference*. IEEE, Dubai.

Mitola, J. and Maguire, G.Q. (1999). Cognitive radio: Making software radios more personal. *IEEE Personal Communications*, 6(4), 13–18.

Mitra, S., Jana, B. and Poray, J. (2016). A novel scheme to detect and remove black hole attack in cognitive radio vehicular ad-hoc networks (CR-VANETs). *International Conference on Computer, Electrical & Communication Engineering*. ICCECE, Kolkata.

Niyato, D., Hossain, E., and Wang, P. (2011). Optimal channel access management with QoS support for cognitive vehicular networks. *IEEE Transactions on Mobile Computing*, 10(4), 573–591.

Ouattara, D. (2014). Apport des réseaux intelligents aux usages et pratiques en e-santé: Une architecture flexible basée sur la technologie radio cognitive pour un suivi efficace et temps réel des patients. PhD Thesis, University of Bordeaux, Bordeaux.

Palicot, J. (2011). *Radio Engineering: From Software Radio to Cognitive Radio*. ISTE Ltd, London and John Wiley & Sons, New York.

Riggio, R., Bradai, A., Harutyunyan, D., Rasheed, T., and Ahmed, T. (2016). Scheduling wireless virtual networks functions. *IEEE Transactions on Network and Service Management*. IEEE publisher, 13(2), 240–252.

Shladover, S.E., Lappin, J., Denaro, R.P., and Smith, B.W. (2014). Introduction: The Transportation Research Board's 2013 Workshop on Road Vehicle Automation. In *Road Vehicle Automation*, Meyer, G. and Beiker, S. (eds). Springer, Cham.

Singh, K.D., Rawat, P., and Bonnin, J.-M. (2014). Cognitive radio for vehicular ad hoc networks (CR-VANETs): Approaches and challenges. *EURASIP Journal on Wireless Communications and Networking*. Springer, 2014(1), 1–22.

Tian, D., Zhou, J., Wang, Y., Sheng, Z., Duan, X., and Leung, V.C.M. (2019). Channel access optimization with adaptive congestion pricing for cognitive vehicular networks: An evolutionary game approach. *IEEE Transactions on Mobile Computing*. IEEE publisher, 19(4), 803–820.

Usha, M. and Ramakrishnan, B. (2019). An enhanced MPR OLSR protocol for efficient node selection process in cognitive radio based VANET. *Wireless Personal Communications*. Springer, 106(2), 763–787.

Wei, Z., Yu, F.R., Tang, H., Liang, C., and Yan, Q. (2016). Securing cognitive radio vehicular Ad hoc networks with trusted lightweight cloud computing. *IEEE Conference on Communication and Network Security*. IEEE, Philadelphia.

List of Authors

Asma AMRAOUI
Abou Bekr Belkaid University
Tlemcen
Algeria

Hasnaâ ANISS
IFSTTAR
Bordeaux
France

Mohammed Anis BENBLIDIA
University of Technology of Troyes
France

Fayssal BENDAOUD
ESI-SBA
Sidi Bel Abbès
Algeria

Badr BENMAMMAR
Abou Bekr Belkaid University
Tlemcen
Algeria

Marion BERBINEAU
IFSTTAR
Bordeaux
France

Bouziane BRIK
CESI
Rouen
France

Mohamed Aymen CHALOUF
IRISA
Rennes
France

Moez ESSEGHIR
University of Technology of Troyes
France

Omessaad HAMDI
IMT Atlantique
Rennes
France

Francine KRIEF
ENSEIRB-MATMECA
Bordeaux
France

Killian LE PAGE
ALTEN
Boulogne-Billancourt
France

Maïssa MBAYE
Gaston Berger University
Saint-Louis
Senegal

Léo MENDIBOURE
LaBRI
Bordeaux
France

Leila MERGHEM-BOULAHIA
University of Technology of Troyes
France

Zeinab MOVAHEDI
Iran University of Science
and Technology
Tehran
Iran

Abderrazaq SEMMOUD
Abou Bekr Belkaid University
Tlemcen
Algeria

Ahmed Khalid Yassine SETTOUTI
Abou Bekr Belkaid University
Tlemcen
Algeria

Index

E, F

G, H

I, K

L, M

N, O

P, Q

R, S

T, V

W, Z